49 Battery-Powered One-IC Projects

Delton T. Horn

FIRST EDITION
FIRST PRINTING

Copyright © 1989 by TAB BOOKS Inc.
Printed in the United States of America

Reproduction or publication of the content in any manner, without express permission of the publisher, is prohibited. The publisher takes no responsibility for the use of any of the materials or methods described in this book, or for the products thereof.

Library of Congress Cataloging-in-Publication Data

Horn, Delton T.
    49 battery-powered one-IC projects / by Delton T. Horn.
      p. cm.
    Includes index.
    ISBN 0-8306-9155-3   ISBN 0-8306-3155-0 (pbk.)
    1. Integrated circuits—Amateurs' manuals. I. Title. II. Title:
Forty-nine battery-powered one-IC projects.
TK9966.H65   1988
621.381'5—dc20   89-34729
                                                                           CIP

TAB BOOKS Inc. offers software for sale. For information and a catalog, please contact TAB Software Department, Blue Ridge Summit, PA 17294-0850.

Questions regarding the content of this book should be addressed to:

    Reader Inquiry Branch
    TAB BOOKS Inc.
    Blue Ridge Summit, PA 17294-0214

Acquisitions Editor: Roland S. Phelps
Technical Editor: Daniel Early
Production: Katherine Brown

Contents

Introduction vi

1 Introduction to the Projects 1
Finding the Components 1
Making Substitutions 5
Breadboarding 11
Construction Techniques 13
Precautions 14

2 Switching Circuits 17
Touch Switch 17
Light-activated Relay 19
Alternate Light-activated Relay 21
VOX Operated Relay 21
Telephone-activated Relay 24
Automated Guest Greeter 25
Schmitt Trigger 28
Switch Debouncer 31
Programmable Timer 35
24-hour Timer 38
Long-duration Timer 41
Timed Touch Switch 46

3 Amplifiers — 48
Audio Amplifier 49
Ceramic Phono Amplifier 54
Audio Mixer 55
Signal Splitter 55
Digital Linear Amplifier 57
Difference Amplifier 59
Logarithmic Amplifier 62

4 Oscillators and Signal Generators — 68
Op Amp Square-wave Generator 68
Triangular-wave Generator 76
Op-Amp Sine-wave Oscillator 76
Digital Sine-wave Generator 82
Digital Phase-shift Oscillator 88
Audio Function Generator 89
Sound-pocket Generator 91
Tone-burst Generator 93

5 Test Equipment and Measuring Devices — 96
Audible Continuity Tester 96
Bargraph 98
Voltage Comparator 100
Voltage Range Detector 103
Logic Probe 107

6 Alarm and Indicator Circuits — 112
Simple Burglar Alarm 112
Flooding Alarm 115
Light-on Alarm 117
Light-off Alarm 118
Light-range Detector 118

7 Filters — 122
Integrator 123
Active Band-pass Filter 125
Wide-band Filter 129
Notch Filter 131
State Variable Filter 135

8 Miscellaneous Projects — 136

LED Flasher 136
Dual LED Flasher 137
Frequency Doubler 139
Sine-to-Square-wave Converter 141
AM Transmitter 143
Binary-to-Hexadecimal Converter 146
Bonus: D/A Converter 149
Bonus: Frequency Halver 153
Bonus: Divide-by-Three Circuit 155

Index — 159

Introduction

Electronics is becoming increasingly complex and sophisticated, even for hobbyists. At one time, believe it or not, an AM radio or a light dimmer was considered to be an advanced project. Today hobbyists are building robots, computers, and high-grade stereo and video equipment.

But there is still a place for simpler projects, and that's where this book comes in. It's nice to be able to turn out a complete, working project in just an hour or two. Beginners should learn their skills on relatively simple and inexpensive projects before tackling more complex projects. Even an advanced hobbyist can enjoy the relaxation of whipping together a quick and easy project. In addition, many of the projects in this book are fun or useful in themselves.

Each project in this book is built around just a single, readily available integrated circuit (IC). None of the projects should cost you more than about $10. Construction time for any of the projects should be well under two hours.

I hope you enjoy these simple, but interesting one-IC projects.

1
Introduction to the Projects

All the projects in this book are relatively simple. Any one of them can easily be constructed in a single evening by either an experienced hobbyist or a beginner. Each project is built around a single integrated circuit, or IC, and a handful of external components.

This introductory chapter will probably be a review for many readers; but even if you have built hundreds of electronic projects, you should still read it just in case there is something you have forgotten or maybe never even thought of. This chapter will offer a number of tips and shortcuts for building any of the projects in this book, or virtually any other electronic project you may come across.

FINDING THE COMPONENTS

You should have no problem in finding any of the components for the projects in this book. In designing and choosing these projects, I have limited myself as much as possible to readily available parts.

Most of the parts can be purchased at a Radio Shack store—and they're everywhere these days. Most cities also have one or more independent parts supply houses.

Another good group of sources for electronic parts is the mail-order companies that advertise in the back of electronics hobbyist magazines. You can find such magazines at almost any news stand. Of the mail-order companies, surplus houses frequently have the best bargains, if they happen to have what you need.

Many mail-order companies have minimum order requirements. If your order adds up to less than a specified amount, you will be charged an extra (and usually hefty) handling fee. Presumably, these companies don't want to be bothered with filling small, less profitable orders. Personally, I feel this is an unfair business practice, since they are already charging a shipping and handling fee on all orders. The minimum order fee is simply penalizing customers for not spending enough to suit the seller.

Some companies have a fairly reasonable minimum order of $5 to $10. Usually there is a base shipping and handling charge, which is increased as the size of the order is increased (usually determined by weight, but sometimes by cost). If you order no more than one or two parts at a time, you will just be wasting money on shipping and handling charges and postage even without the minimum order fee.

Incidentally, you must take the shipping and handling charges into account when determining which mail-order company offers the best bargains. You can't go by their price lists, which are often misleading. If company A charges an average of 2 to 10 cents per component more than company B, but has a lower shipping and handling charge, you may get more for your money from company A.

A few companies have really ridiculous minimum order requirements: $20 or more. At least one outfit actually has a minimum order of $50. These companies obviously don't want to deal with hobbyists or small businesses, so it is probably best to avoid them altogether. If there is any problem, they probably won't be too interested in giving you very good service, unless you are placing an industrial sized order.

It usually isn't too hard to make up a minimum order of five or ten dollars. You can usually make the minimum by ordering components for several projects at once. Or, you can combine orders with a friend or two.

You can also sometimes reach the minimum order level and occasionally find some great bargains by ordering grab bags of components. A grab bag is a "pig in the poke" assortment of parts. Be careful though: don't spend too much for grab bags at any one company unless you know them and the kind of quality they offer.

Grab bags are always sold "as is." Some companies offer grab bags that are almost always worth more than the price. Others tend to give you a bag of useless junk. In some cases, a grab bag is literally floor sweepings from some factory. You get components that have dropped on the floor (and perhaps have been crushed under foot), along with who knows what else. Once I actually got part of a very, very old sandwich. The reliability and honesty of the seller is a prime consideration when you are buying grab bags. Parts in grab bags are also usually untested, so expect to get some duds.

Generally the best and most reliable bargains are to be had with classified grab bags. Rather than a general grab bag of who knows what, get a grab bag of 100 resistors, or 50 capacitors, or something similar. Sure, you'll probably get some pretty oddball values, but you'll almost certainly get some items you can use. It isn't always possible, but try to stick with grab bags that are guaranteed to include marked components. Figuring out the values of a hundred or so unmarked components is usually more trouble than it's worth.

Grab bags are a good way to build up your "junk box." Every electronics hobbyist should have a "junk box." This is a collection of common components that are likely to be useful in future projects. This cuts down the costs of your hobby considerably. Experienced hobbyists with well stocked junk boxes will probably be able to build many of the projects in this book without spending a cent.

Another way to build up your junk box is by acquiring and dismantling discarded equipment. If your next-door neighbor is throwing out a broken radio, ask him for it. Or, if you're willing to face some possible embarrassment, sneak it out of his trash. Even if it doesn't work at all, it is still bound to have a number of perfectly good parts. When you "cannibalize" old equipment, concentrate on the more expensive components like switches, semiconductors, and potentiometers. Don't bother with less expensive parts like resistors and disc capacitors. Removing them from the circuit will be more trouble than it's worth, and the leads are likely to be too short to be reusable. If the equipment is very old, leave the electrolytic capacitors; the dielectric might be dried out.

Incidentally, be very careful around any large capacitors. Even if no power has been applied to the circuit for quite a while, they might still hold a considerable charge. If you're not careful you could receive a painful, and possibly dangerous, electrical shock.

Try not to put any defective components into your junk box. They'll only cause you grief later. Obviously, if you use a defective component, your project won't work properly. If you have any doubt about the quality of a component and you don't have a suitable tester available, discard the component.

You should also save nonelectronic parts like knobs, plugs, jacks, and line cords. If there is any chance of reusing the cabinet, save it. Cabinets for projects tend to be rather expensive. If you can live with a few inappropriate markings or a few excess holes for the original controls, reusing an old cabinet will be a worthwhile saving.

You can also replenish your junk box by dismantling your old projects that you're no longer using. If you made a mistake on a project and it didn't work, it still probably has some perfectly good, reusable components. Don't throw them away.

Keep your junk box well organized. If you throw everything into a shoebox, it probably won't do you much good. You'll never be able to find the part you need when you need it. (But somehow it will always manage to turn up in your junk box right after you've ordered a new one.)

Get a multidrawer parts cabinet, or make some kind of homemade dividers. Separate your junk box components by type. If you have a large number of a given type of component, such as resistors and capacitors, you'll also want to subdivide them by value. You don't need a separate container for every individual value. Organize these parts by a value range. For example, resistors might be sorted in the following categories:

<100 ohms
100 ohms–1K
1K–10K
10K–100K
100K–1 Megohm
>1 Megohm

Your storage compartments don't have to be elaborate or expensive. I have stored small parts in ordinary business envelopes. This proved handy, because the type of part enclosed and the value, or value range, could be marked directly on the envelope.

MAKING SUBSTITUTIONS

Occasionally you won't have exactly the right part on hand for a project. In some projects, certain components are very critical, but usually substitutions can be made. The projects in this book are fairly flexible, and you shouldn't have any problems making substitutions. In many cases, you are encouraged to experiment with other component values. If a value does happen to be critical, this point will be mentioned in the text for that project.

Always make sure that any substitute component has voltage, current, and power ratings equal to or greater than those of the part being replaced. If a ¼-watt resistor is called for (as in most of the projects in this book), you can certainly use a ½-watt resistor. It will be slightly larger physically, but that probably won't be important in most hobbyist applications. On the other hand, you should not replace a ½-watt resistor with a ¼-watt unit. It might not be able to stand up to the power fed through it by the circuit. If a resistor burns out or changes value due to overheating, your project won't work properly.

Similarly, if a 100 μF, 15-volt electrolytic capacitor is called for, feel free to use a 100 μF 25-volt device. If the original diode has a PIV rating of 100 volts, you'll have no problem using a diode with a PIV rating of 200, or even 300 volts.

These power ratings are minimum values for the component used in the circuit. As a rule, you should use a component close to the called for ratings. Components with higher ratings tend to be physically larger, and—more important to the hobbyist—more expensive.

There is one important exception to this general rule of overrating replacement components: fuses. *Never* replace any fuse with one that has a *higher* current rating. The current rating for a fuse is *not* selected arbitrarily. The fuse is used to protect the other components in the circuit. It is rated for the maximum amount of current that can be safely drawn by the circuit. If you use a larger value fuse, an expensive transistor or IC could blow out to protect the fuse, which pretty much defeats the whole point.

To substitute semiconductors (transistors and ICs) you will need a suitable substitution guide. These are available from various publishers and component manufacturers. It is best to have a number of substitution guides from a variety of sources. Many manufacturers market a line of components specifically designed as general substitutes. For example, there is the Motorola HEP series, and the Sylvania ECG line.

A general purpose substitute will work in most of the projects in this book. If the exact component is required, this will be noted in the text. To the best of my knowledge, there are no substitutes available for some of the ICs used in these projects.

When using a substitute component, always be sure to double-check pin numbering for ICs or the lead arrangement for transistors. Sometimes a suitable electrical substitute won't be an exact mechanical substitute. In some cases you might have to modify the circuit wiring slightly.

Most substitutions will be made with passive components, such as resistors and capacitors. If you don't have the exact value called for, you may have something in your junk box that is close enough to do the job.

Most passive component values are pretty flexible within a specific range or tolerance. For example, resistors are available with the following tolerance ratings:

20%
10%
5%
1%
0.1%

The actual value of the component may be off from the stated value up to the tolerance percentage of the stated value. As an example, let's assume we have a 100K (100,000 ohm) resistor at each of the tolerances just listed. The maximum error, and the range of values for each tolerance are given in Table 1-1.

Table 1-1. Resistor Values.

Rating	Maximum Error	Minimum Actual Value	Maximum Actual Value
20%	±20000	80,000	120,000
10%	±10000	90,000	110,000
5%	±5000	95,000	105,000
1%	±1000	99,000	101,000
0.1%	±100	99,900	100,100

Notice that even a 20 percent tolerance resistor may be exactly at its stated value. The tolerance rating simply specifies the *maximum* error range. The component is guaranteed to be somewhere within the range defined by the tolerance rating.

Generally, 20 percent tolerance resistors are only used in the most non-critical applications. They are becoming increasingly rare.

Very tight tolerance resistors, such as 1 percent or 0.1 percent units, tend to be relatively expensive, so they are used only in those applications where very high precision is mandatory.

For most applications, 5 percent or 10 percent tolerance resistors will be your best bet. This is true for all the projects in this book. Unless noted otherwise, you may use either 5 percent or 10 percent tolerance resistors in these projects.

You can always substitute a component with a tighter tolerance rating than the one called for in the parts list. For example, if a 10 percent unit is called for, you can certainly use a 5 percent device.

You can go in the other direction too, if you use an ohmmeter to determine the actual value of the resistor. You can use a 20 percent resistor in place of a 10 percent unit, if the measured value is no more than 10 percent off from the nominal, stated value.

If you don't have the exact stated value handy, you can often substitute a resistor with a value that is close. For example, if the project calls for a 4.7K 10 percent resistor, you could probably get away with using a 3.9K resistor, especially if it had a 5% tolerance. To be on the safe side, you should breadboard the circuit with such a substitution to make sure it works properly before permanently soldering the components into place.

You can also make up unavailable values (including unusual values) by combining resistors in series and in parallel. You don't have to use a single component to give the desired resistance.

For resistors in series, as shown in Fig. 1-1, the resistances are simply added together:

$$Rt = R1 + R2 + \ldots Rn$$

For example, let's say we have three 100 ohm resistors in series. The total effective resistance will be:

$$\begin{aligned}Rt &= 100 + 100 + 100 \\ &= 300 \text{ ohms}\end{aligned}$$

Fig. 1-1. Resistances in series add.

For parallel combinations, like the one illustrated in Fig. 1-2, the reciprocal of the total is equal to the sum of the reciprocals of each of the component resistances. This sounds a lot harder than it really is. It is much clearer in equation form:

$$1/Rt = 1/R1 + 1/R2 + \ldots 1/Rn$$

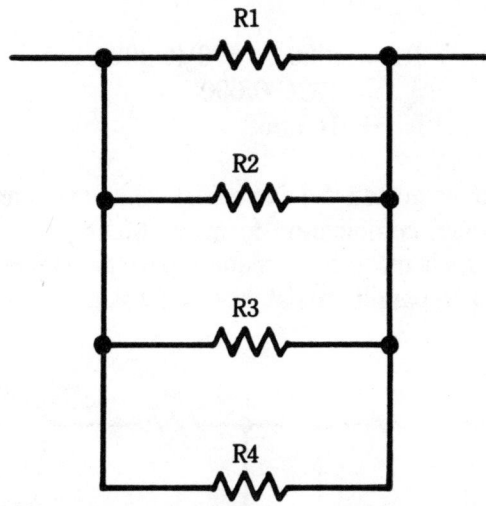

Fig. 1-2. When resistors are combined in parallel, the total effective resistance is less than any of the component resistances.

As an example, let's assume we have three 100 ohm resistors in parallel. The total effective resistance works out to:

$$\begin{aligned}
1/Rt &= 1/100 + 1/100 + 1/100 \\
&= 0.01 + 0.01 + 0.01 \\
&= 0.03 \\
Rt &= 1/0.03 \\
&= 33.3333 \text{ ohms}
\end{aligned}$$

Notice that for series combinations, the total effective resistance is always *greater* than any of the component resistances. On the other hand, for parallel combinations, the total effective resistance is always *smaller* than any of the component resistances.

There is a modified form of the parallel equation that can be used when there are just two resistances in parallel:

$$Rt = (R1 \times R2)/(R1 + R2)$$

For example, let's assume we have a 220 ohm resistor in parallel with a 470 ohm resistor. In this case, the total effective resistance is:

$$\begin{aligned} Rt &= (220 \times 470)/(220 + 470) \\ &= 103400/690 \\ &= 149.855 \end{aligned}$$

We would have gotten the same final value if we had used the regular parallel combination formula. Most people find this modified formula more convenient for two parallel resistances.

Series and parallel resistances may also be combined, as shown in Fig. 1-3.

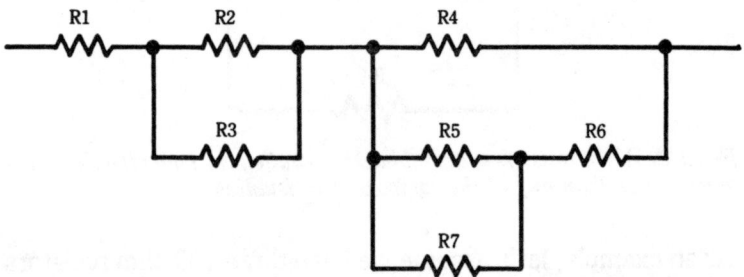

Fig. 1-3. Series and parallel resistances can be combined in a single circuit.

Capacitors can be substituted in a manner similar to resistors. Capacitor tolerances, especially for large electrolytic capacitors, tend to be much wider than resistor tolerances, so you have much more room for making substitutions. If the stated value is reasonably close, it will probably work fine, unless a high precision capacitor is specifically called for.

There is some variation in capacitor marking, so components which are essentially identical may be marked differently by different manufacturers. For example, there is rarely any noticeable difference between disc capacitors marked 0.2 μF, or 0.22 μF, or even 0.25 μF. In many applications, you could even substitute a capacitor marked 0.15 μF.

Capacitors may be combined in series or parallel, just like resistors. But the rules for determining the total effective value are the exact opposite of those used with resistors. The formula for capacitors in series is:

$$1/C_t = 1/C_1 + 1/C_2 + \ldots 1/C_x$$

While for parallel capacitances, the formula is:

$$C_t = C_1 + C_2 + \ldots C_x$$

Because of the wider tolerances of capacitors, the equations are much less precise for capacitors. If you can get the total effective capacitance reasonably close to the desired value, it will probably be perfectly acceptable. Series and parallel capacitances may also be combined.

BREADBOARDING

In the early days of electronics, experimenters mounted components on a piece of wood, or a breadboard. The term "breadboarding" has survived to describe any form of solderless temporary construction of an electronic circuit.

Today breadboarding is generally done on a special solderless socket, like the one shown in Fig. 1-4. The socket is covered with rows of holes that component leads can fit into. The holes are spaced to accommodate the pins of ICs in DIP type housings.

Within the socket, the holes are internally connected in a specific pattern. A common interconnection pattern is illustrated in Fig. 1-5.

Full integrated breadboarding systems are also available. In addition to the solderless socket, the system includes

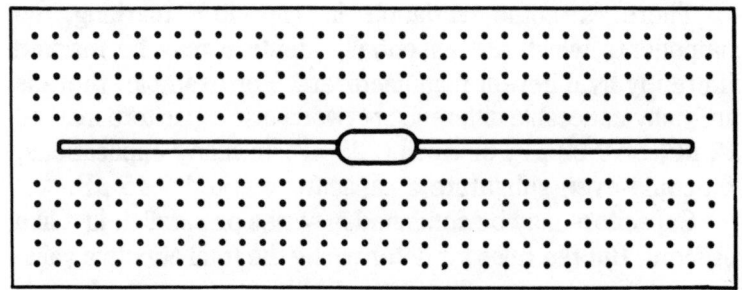

Fig. 1-4. Breadboarding is usually done on a solderless socket.

commonly used circuit modules, such as variable power supplies and oscillators. You won't have to build these auxiliary circuits each time they are needed. Conveniently mounted switches and potentiometers are also provided.

It is usually a good idea to breadboard any circuit before constructing it permanently. You might decide the circuit doesn't really do what you want it to do, so you might as well reuse the components in some other project. Any problems are easier to correct in a breadboarded circuit. Suppose you happen to have a dud IC. It is easy to replace it in a solderless socket, but it would be a nuisance to change it if you had to desolder it and solder in a new one.

Occasionally an error may creep into a schematic diagram, despite the best efforts of editors and technical writers. If you breadboard the circuit first, you will find out that there is a problem and it will be a lot easier to find and correct the erroneous or missing connection.

Best of all, breadboarding permits you to experiment with, and easily modify the circuit. You can make sure any non-identical substitution works. You might want to see how circuit performance is changed if this resistance is increased, or if that capacitance is decreased. Perhaps you might want to add another output device. You'll need to find out if this addition will load down the circuit excessively. If you breadboard the project first, you can find out the answers to all such questions before you have to commit yourself with the soldering iron.

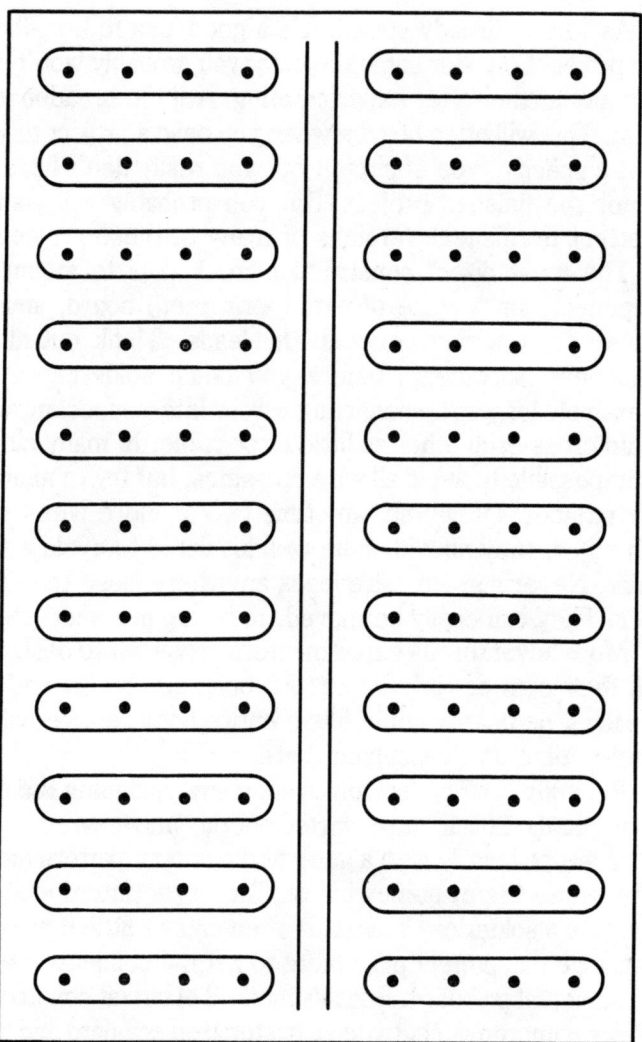

Fig. 1-5. This is a common interconnection pattern for a solderless socket.

CONSTRUCTION TECHNIQUES

There are several different construction techniques you may use. They are all suitable for most of the projects presented in this book. It really boils down to a matter of personal preference.

As I have already stated, it's a good idea to breadboard each project first. For some projects, you probably won't want to go any further after experimenting with the breadboarded circuit. This will often be true when you build a project to learn about a specific type of circuit, but you really don't have any use for the finished project. But you probably will want to construct permanent versions of many of these projects.

The most direct construction method is to mount the components on a piece of perf (perforated) board, and use point-to-point wiring between the leads. Think out all the component placements before you begin soldering. Avoid excessively long interconnecting wires. Interconnecting wires should cross each other as little as possible. In many circuits it is impossible to avoid all wire crossings, but try to minimize their number. Of course, any time two or more wires cross each other, they should all be well insulated to avoid a short circuit. Never run any bare leads anywhere close to one another. They can easily be moved, resulting in a short circuit.

More adventurous experimenters may want to design and etch their own pc (printed circuit) boards. The ins and outs of making pc boards could fill an entire book, so we will not go into detail on that subject here.

Recently, many electronics suppliers, including the ever-present Radio Shack, have started offering universal pc boards. These are pc boards with a generalized copper pattern already etched onto them, somewhat similar to the interconnection pattern of a solderless socket. It often takes a little more work to arrange component placement to get the connections right on a universal pc board. But this method of circuit construction is a nice compromise between a customized pc board and point-to-point perf board construction.

PRECAUTIONS

With any type of circuit construction, you should consider using sockets for the ICs. A socket is a relatively cheap form of insurance. An IC is a semiconductor, of course, and is therefore heat sensitive. Soldering all of those closely spaced pins could easily result in the chip being over-heated and damaged. If you do solder an IC directly into a circuit, the use of a soldering heat sink is practically mandatory.

Some people in the field feel IC sockets are usually wasteful. In many cases, a socket may cost more than the IC being protected. This is true, but if you do damage an IC in direct soldering (or get a bad one—it happens occasionally), you will have to desolder it, then solder in a replacement. That is a lot of very tedious extra work. So a socket can save you quite a bit of frustration.

Always remember never to make *any* changes in a circuit with the power connected. This especially includes putting an IC into a socket or taking it out. *Always* remove power first.

In some very high frequency circuits, a socket may interfere with proper operation of the circuit, especially if a cheap socket is used. However, this book does not include any such "fussy" circuits. Unless you do a lot of work in the upper radio bands, or in high-speed digital systems, you will probably never run into such problems.

Whether or not you are using sockets, be very careful with the orientation of all ICs. Don't install an IC backwards. Also, you must make sure none of the pins become bent under the body of the IC, where they won't make electrical contact with the circuit. Make it a habit to always double-check such potential problems before applying power to any project. To help in their orientation, most ICs have a clear marking on the front end, or an indication of the location of pin No. 1, as illustrated in Fig. 1-6.

When soldering, whether an IC, or an IC socket, always be aware of how closely spaced adjacent pins are to one another. It is very easy to inadvertently create a solder bridge between two or more pins if you're not very, very careful. This is something else you should always double-check before applying power to the circuit.

Any specialized precautions for specific projects will be described in the text, where appropriate.

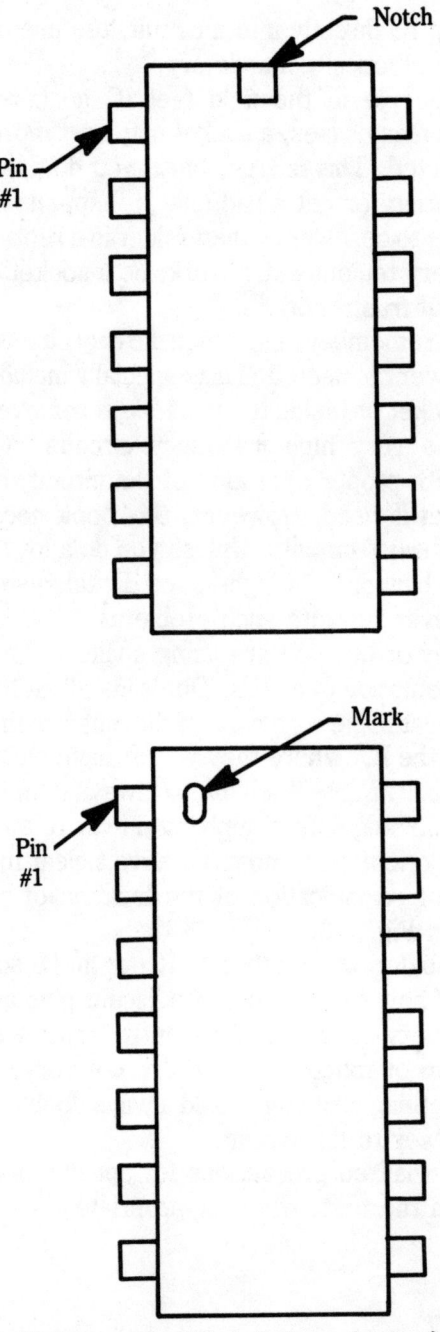

Fig. 1-6. Most ICs have a marking to help you locate pin No. 1.

2
Switching Circuits

Probably the most basic of all electronic functions is switching. A mechanical switch is certainly a mundane device. It simply turns something on or off when a mechanical part is physically moved.

But many more sophisticated switching applications are possible, including automated and timed switches. This chapter features a number of projects built around relays, sensors, timers, and other switching devices.

Many of these projects can be easily modified to be included in some very sophisticated control systems. Others, though very simple projects, are quite powerful and versatile, even without any modification.

PROJECT 1: TOUCH SWITCH

With a touch switch, you can control almost anything with the lightest touch. A touch switch may be activated with a fingertip, or, if your hands are full, an elbow or a foot. If you were so inclined, you could even operate it with your nose. It can also be used in many hidden sensor (alarm) applications. Besides, the light touch operation of a touch switch is fun and intriguing on its own.

A simple, but practical touch switch circuit is shown in Fig. 2-1. The parts list for this project is given in Table 2-1.

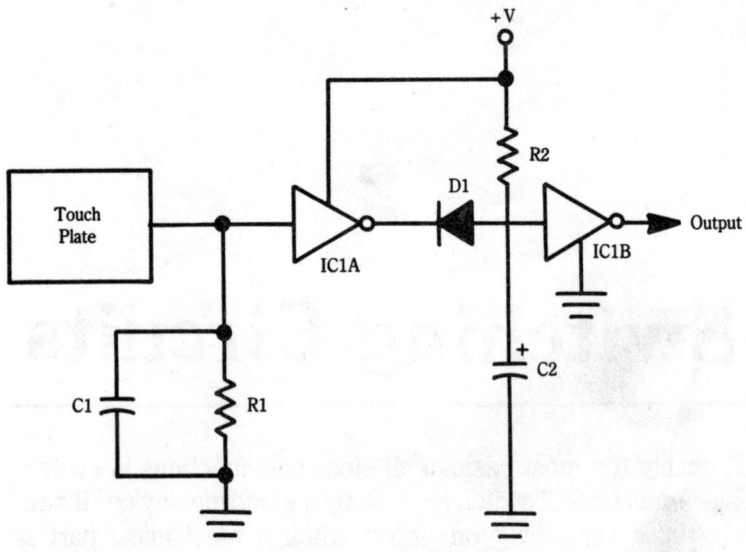

Fig. 2-1. This is a simple but practical touch switch circuit. Project 1

Table 2-1. Parts List for the Touch Switch Project of Fig. 2-1.

Schematic Label	Part
IC1	CD4049 hex inverter
D1	1N914 diode
C1	100 pF capacitor
C2	1 μF capacitor
R1	10 Megohm, ¼-watt resistor
R2	100K, ¼-watt resistor

The touch plate is an exposed metallic contact. A small piece of unetched copper-clad PC board will do just fine as the touch plate. The plate's dimensions will be influenced by the desired application. For most purposes, the plate should probably be about one inch square.

Obviously, the touch plate *must* be *100 percent isolated* from *any* ac power source for safety. The person operating this project comes in direct contact with a conductor (the touch plate) that is wired into the circuit. If any ac voltage manages to get through to the touch plate, the operator could suffer a painful, and quite possibly dangerous electrical shock. Only low power dc voltages should be allowed to flow through the touch switch circuit. Isolate the control circuit from any ac load through a relay or an optoisolator. The touch switch circuit should be operated off battery power *only*. An ac-to-dc converter type power supply could conceivably fail and feed 120 volts ac into the circuit and the touch plate. It's better to be too safe than sorry.

I strongly recommend that you breadboard this circuit before constructing a permanent version. While the component values are not especially critical, sometimes this type of circuit may be a bit fussy. You might need to make some minor changes or do some fine tuning before the project operates reliably. You might want to replace the two fixed resistors with trimpots to permit fine tuning.

This circuit, and most other touch switch circuits, takes advantage of the fact that 60 Hz power signals are almost always nearby. Low level 60 Hz signals from power lines are picked up by the body, which acts as an antenna, and transmitted through a fingertip or other body part to a small touch plate. This input signal triggers the control circuit built around the two digital inverter stages, and the output goes High.

PROJECT 2: LIGHT-ACTIVATED RELAY

This project permits fully automated control over virtually any electrically powered device. This switching circuit is controlled by ambient light levels.

The circuit, which is shown in Fig. 2-2, features a fully adjustable sensitivity control. The parts list for this project is given in Table 2-2.

This circuit isn't terribly complex. The op amp (IC1) is set up as a voltage comparator. The voltage dropped across

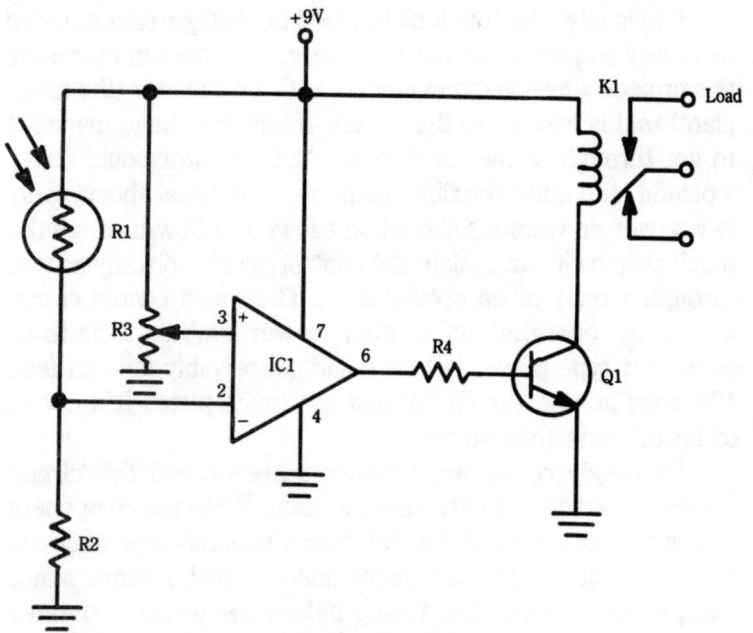

Fig. 2-2. A relay can be activated by light striking a sensor. Project 2

**Table 2-2. Parts List for the
Light-activated Relay Project of Fig. 2-2.**

Schematic Label	Part
IC1	Op amp (741, or similar)
Q1	NPN transistor (2N2222, or similar)
K1	9 volt relay contacts to suit load
R1	photoresistor
R2, R4	100K, ¼-watt resistor
R3	250K potentiometer

the photoresistor (R1) is compared to the voltage dropped across the upper half of the pontentiometer (R3). Adjusting potentiometer R3 sets the sensitivity of the light sensor. That is, the setting of R3 determines how much light is needed to trigger the circuit.

If there is too little light, the reference voltage set by R3 will be greater than the variable voltage determined by R1. The more light there is striking the sensor, the higher the variable voltage becomes, while the reference voltage remains constant. When the comparator switches over, it triggers the relay, which controls the load. A latching relay might be used in certain applications. In other applications, you will want the relay to become deactivated when the light level drops below the trigger point. If a latching relay is used, one pulse of light will activate the relay and a second pulse of light will deactivate it.

The transistor (Q1) is used to amplify the comparator's output signal. A low power relay is not needed. The transistor can drive a moderately large relay. Almost any NPN transistor may be used for Q1.

PROJECT 3: ALTERNATE LIGHT-ACTIVATED RELAY

This light activated relay project is similar to Project 2, but uses a completely different approach to the circuitry.

This version, shown in Fig. 2-3, is built around a 555 timer. Notice that this IC is not wired in one of its common configurations.

R2, the photoresistor is the light sensor. Potentiometer R1 controls the sensitivity of the circuit. When the light level shining on the photoresistor exceeds a specific level, the 555 timer is triggered, activating the relay.

The parts list for this project is given in Table 2-3.

PROJECT 4: VOX RELAY

Have you ever had your hands full when you needed to turn something on or off? Do you like the convenience of automation? Do you enjoy impressing your friends?

If you answered yes to any of those questions, then this project is ideal for you. It is a VOX relay. *vox* means "voice operated." In this project, a spoken command or other noise activates the contacts of a relay after a brief delay.

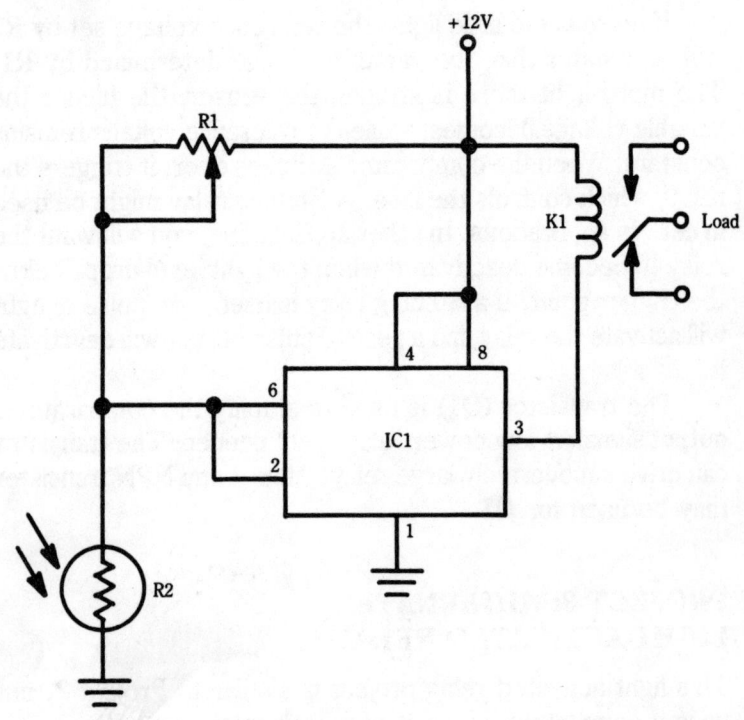

Fig. 2-3. This alternate light activated relay circuit uses a 555 timer. Project 3

Table 2-3. Parts List for the Alternate Light-activated Relay Project of Fig. 2-3.

Schematic Label	Part
IC1	555 timer
K1	12 Volt relay (contacts to suit load)
R1	10K potentiometer
R2	photoresistor

The circuit for this project is shown in Fig. 2-4. The parts list is given in Table 2-4.

Almost any small relay can be activated with this circuit, permitting it to control almost any small to moderate load. If you need to control a larger load, the small relay can be used to control a second, larger relay, as illustrated in Fig. 2-5.

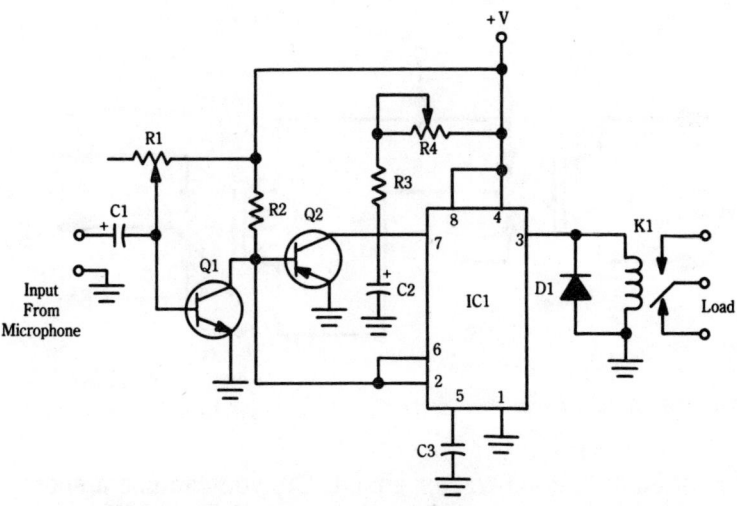

Fig. 2-4. A sharp sound activates the relay in this circuit. Project 4

Table 2-4. Parts List for the VOX Relay Project of Fig. 2-4.

Schematic Label	Part
IC1	555 timer
Q1	NPN transistor (2N3904, or similar)
Q2	PNP transistor (2N3906, or similar)
D1	1N4002 diode
K1	relay with contacts rated to suit load
C1	1 μF 15V electrolytic capacitor
C2	50 μF 15V electrolytic capacitor
C3	0.01 μF capacitor
R1	10K potentiometer
R2, R3	1K, ¼-watt resistor
R4	100K potentiometer

Potentiometer R1 controls the sensitivity of the circuit. This means you can adjust how loud the triggering sound must be. To some degree, the sensitivity control will allow you to use the VOX relay even in moderately noisy environments.

For best results with this project, the triggering sound should be loud, distinct, and sharp. A good, sharp hand-clap

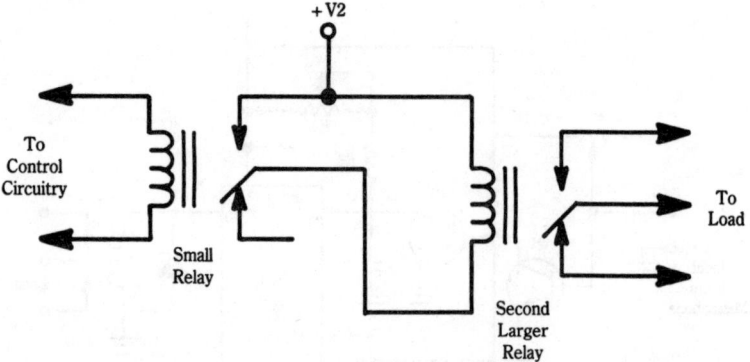

Fig. 2-5. A small relay can drive a second, larger relay.

will often be a good trigger sound. Or, you can use a short, strong cry of "Go!", or "On!", or even "Hey!" The word used doesn't matter, of course. As long as it is short and uttered loudly, clearly and sharply.

The 555 timer introduces a slight delay to the operation of the circuit. The exact amount of delay is determined by the setting of potentiometer R4. With the component values listed in the parts list, the delay may be set anywhere from about 0.05 second to approximately five full seconds. For longer delay settings, a longer triggering sound may be required. A handclap might be too short for reliable triggering. A shout, or whistle would work better under such circumstances.

The advantage of a longer delay time is that it reduces the chances of some stray transient sound accidentally activating the relay at an undesired time. Essentially, the delay time determines the minimum length of the triggering sound.

Because this circuit can be controlled by almost any type of sound, and can control almost anything via the relay, its potential applications are limited only by your imagination.

PROJECT 5: TELEPHONE-ACTIVATED RELAY

The common telephone can be used as a remote control device. When the phone rings, this project activates a relay, controlling almost any load. You can control any electrical device in your absence, simply by phoning home. Of course, there are definite flaws in this simple scheme. If anybody else dials your number, they will also activate the relay. If you're very well off financially, you might be able to afford a second, unlisted number just for remote control purposes, but that solution probably isn't too practical for most of us. Besides, a wrong number can still give unwanted control.

A more practical application would be to use it to drive a remote ringer device, for when you're in the back yard, or somewhere else where you might not hear the phone ring. Alternatively, it can flash a light if you are in a place that is too noisy to let you hear the phone ringing reliably. This would be useful for many audiophiles who like to turn their stereos up very loud.

You could also use this project to turn on a light when the phone rings at night. No more stumbling to the phone in the dark when you get an unexpected late night call. And those late night calls seem to be wrong numbers more or often than not, so a stubbed toe is all the more annoying.

The schematic for this project is given in Fig. 2-6. The parts list appears in Table 2-5.

Notice that this project requires a direct connection to the telephone lines. To prevent possible legal hassles, be sure to check with your local telephone company before connecting anything to their lines. Often, all that is needed is notification of the extra equipment. But if you hook it up without notifying them, you could face some penalties in certain areas of the country.

PROJECT 6: AUTOMATED GUEST GREETER

This one is a very handy project for the homeowner who doesn't want to waste energy.

You probably don't want to leave your porch light on all the time, just in case you might get an unexpected guest after dark. But, on the other hand, if you do get such guests, you don't want them to have to wait outside in the dark until you can answer the door. The circuit shown in Fig. 2-7 shows a clever solution to this problem. The parts list for this project is given in Table 2-6.

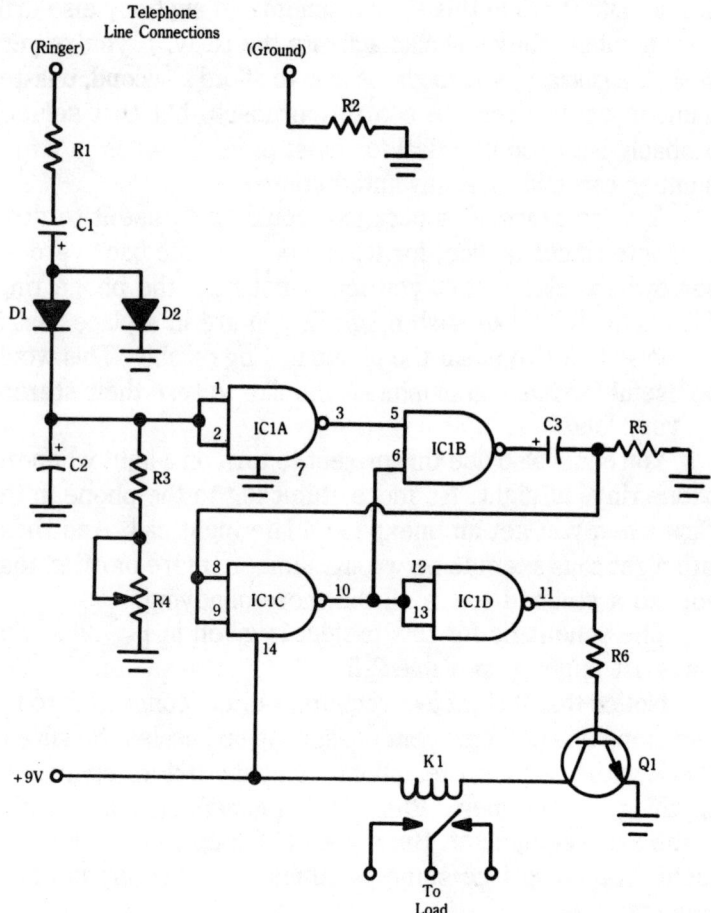

Fig. 2-6. This telephone activated relay circuit can automatically turn on a light when you receive a late night phone call. Project 5

When the doorbell is rung, the porch light will automatically be switched on for a predetermined period of time, then it will be automatically turned off.

Ordinarily, most doorbells are activated with a SPST switch. In this project, we replace the original doorbell switch with a DPST or DPDT switch. One section, or pole, of the switch is connected directly to the doorbell so it functions in the normal manner. The second half (pole) of the switch is

**Table 2-5. Parts List for the
Telephone Activated Relay Project of Fig. 2-6.**

Schematic Label	Part
IC1	CD4011 quad NAND gate
Q1	NPN transistor (2N2222, or similar)
D1, D2	1N4002 diode
K1	relay selected for desired application
C1, C2	1 µF 50V electrolytic capacitor
C3	100 µF 50V electrolytic capacitor
R1, R2	390K, ¼-watt resistor
R3, R5	1 Megohm, ¼-watt resistor
R4	1 Megohm potentiometer
R6	1K, ¼-watt resistor

used to send a trigger signal to a monostable timer (IC1). When triggered, the timer's output goes High, activating the output relay for a period of time determined by resistor R1 and capacitor C1. The formula for the time period is:

$$T = 1.1 R1 C1$$

where T is the time period in seconds, R1 is the resistance in ohms, and C1 is the capacitance in farads.

For guest greeting applications, I think a time period of four or five minutes would probably be best. I suggest these component values:

$$R1 = 1 \text{ Megohm}$$
$$C1 = 250 \text{ µF}$$

This combination gives a time period of 275 seconds, or about 4.6 minutes. Of course, you may choose to substitute other component values for R1 and C1.

Diode D1 prevents back EMF from damaging the relay coil. The relay should be selected so that its contacts are rated

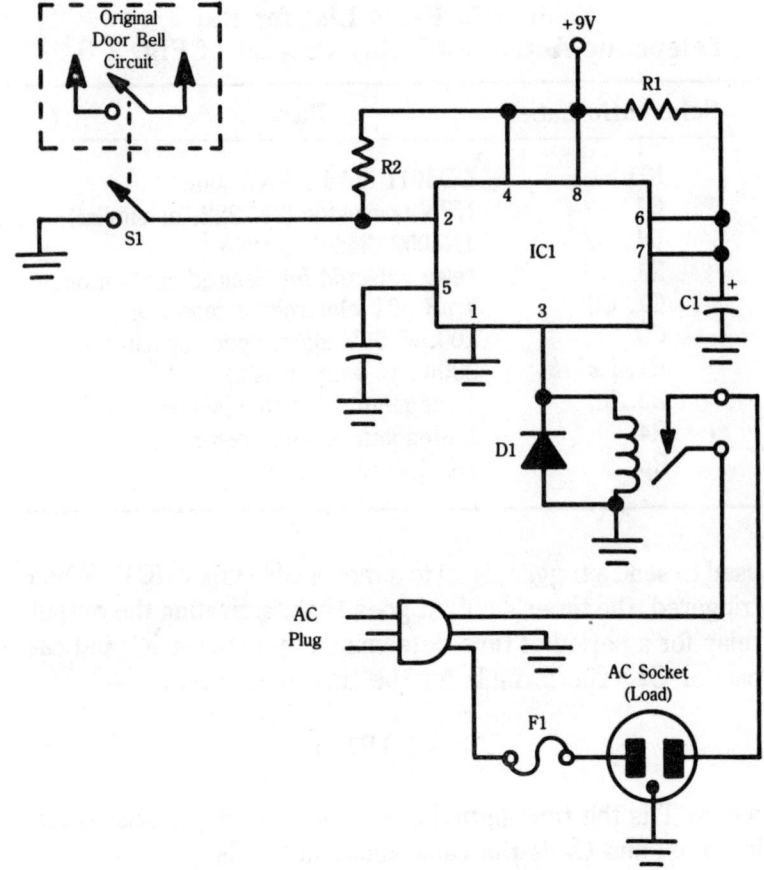

Fig. 2-7. This circuit is an automated guest greeter. Project 6

for slightly more than the intended load. For a light, determining the load is easy—just look at the wattage rating of the bulb. The fuse should also be selected to suit the load. For safety's sake, do not omit the fuse from this circuit.

PROJECT 7: SCHMITT TRIGGER

A Schmitt trigger is used to create clean rectangular wave pulse signals from any input signal. This is especially useful in data transmission applications where noise pick-up can cause confusion and false triggering.

Table 2-6. Parts List for the Automated Guest Greeter Project of Fig. 2-7.

Schematic Label	Part
IC1	555 timer
D1	1N4002
K1	relay to suit load
F1	fuse to suit load
C1	250 µF 35V electrolytic capacitor
C2	0.01 µF capacitor
R1	1 Megohm, ¼-watt resistor
R2	10K, ¼-watt resistor
S1	DPST switch (replacing original SPST doorbell switch)
*	AC socket
*	AC plug

Look at the noisy input signal in Fig. 2-8A. It is not easy to determine just where the desired pulses are located in this mess. A circuit being triggered by these pulses can be confused by the noisy signal. It may trigger when it is supposed to, or it may be falsely triggered by a noise spike where no trigger pulse was intended.

A Schmitt trigger can clean up most noisy signal problems and recover the desired trigger pulses in most cases, as shown in Fig. 2-8B.

A Schmitt trigger circuit is designed to apply a trigger pulse to another circuit when its input signal reaches a specific value. A basic Schmitt trigger circuit is illustrated in Fig. 2-9.

Most practical Schmitt trigger circuits feature controlled hysteresis. This is basically a built-in lag, or grey area between the cut-on and cut-off points. Signals within the hysteresis region are simply ignored by the Schmitt trigger circuit. This makes the circuit more reliable and less prone to false triggering due to noise spikes. The output is switched on by a higher voltage than is used to switch it back off. Minor fluctuations in the input signal do not affect the output signal.

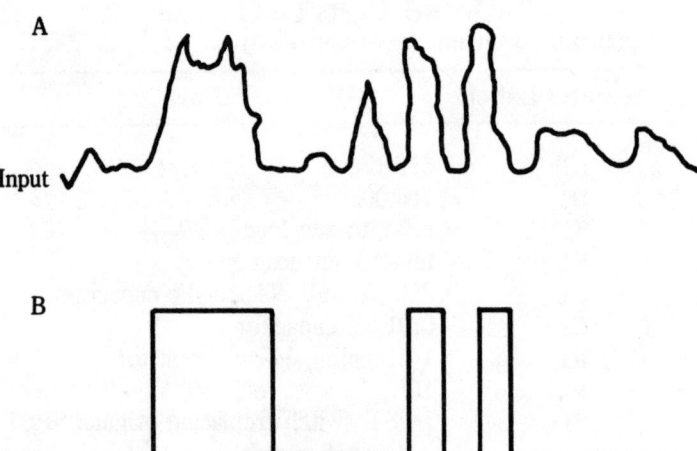

Often the desired pulsewave signals are buried in noise.

Fig. 2-8. Noisy signals can make it difficult for a digital circuit to locate the desired pulses.

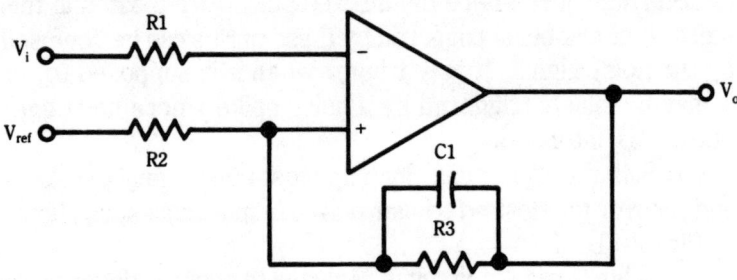

Fig. 2-9. This is a basic Schmitt trigger circuit.

A practical Schmitt trigger circuit is shown in Fig. 2-10. In this circuit it is possible to change the hysteresis without changing the switching voltage level. In some applications, the switch-on or the switch-off voltage level will need to be precisely set by the reference voltage (Vr), and the other switching level is determined by the amount of desired hysteresis. Adjusting potentiometer R2 will change the hysteresis of this circuit.

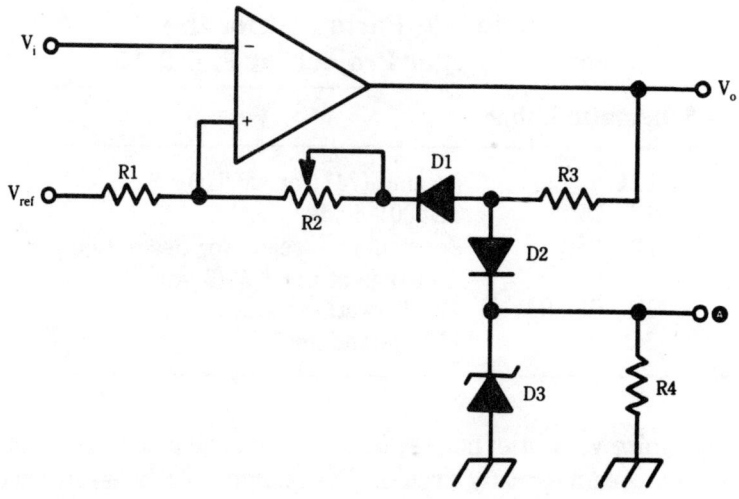

In this Schmitt trigger circuit, the amount of hysteresis is manually controllable.

Fig. 2-10. A practical Schmitt trigger circuit is shown here. Project 7

A fringe benefit of this circuit is that a second output may be tapped off at point A in the schematic. This output signal will switch between ground (0 volts) and the voltage determined by zener diode D3. This switching will be in step with the action of the main Schmitt trigger. The secondary output is useful when the full saturation voltage swing of the op amp is too large for the circuit being triggered by the output pulses.

Besides cleaning up noisy signals, a Schmitt trigger may be used to interface certain analog signals with digital circuitry.

The parts list for this project is given in Table 2-7.

PROJECT 8: SWITCH DEBOUNCER

No mechanical switch is perfect. What is? Rather than neatly making or breaking contact, as shown in Fig. 2-11A, they actually tend to bounce open and shut several times before settling into position, as shown in Fig. 2-11B. This takes only a tiny fraction of a second. For most analog applications, this bouncing rarely matters. It won't even be noticeable in the circuit's operation. Digital circuits, however, are designed to

Table 2-7. Parts List for the Schmitt Trigger Project of Fig. 2-10.

Schematic Label	Part
IC1	op amp (741, or similar)
D1, D2	1N4001 diode
D3	Zener diode—select for desired output voltage at point A (5 volts typical)
R1, R3, R4	1K, ¼-watt resistor
R2	25K potentiometer

recognize very brief pulses, so bouncing switch contacts could be a problem in many circuits. For example, let's assume we want to increment a counter each time a push-button switch is depressed. The bouncing could cause the counter to be incremented several times instead of just once for each time the switch is pushed. Obviously this could be a major problem, possibly even rendering some digital circuits completely useless.

What we need is some way to get the digital circuitry to ignore the unwanted bouncing contact pulses. A good way to do this is to have the first closure of the switch contacts trigger a monostable multivibrator with an output time period that is slightly longer than the bouncing (settle-in) time of the switch.

The circuit for the switch debouncer is shown in Fig. 2-12. The parts list is given in Table 2-8.

As soon as the switch contacts first make contact, the monostable multivibrator, or one-shot timer, is triggered. Further switch openings and closings will be ignored until the multivibrator completes its timing period. This timing period does not have to be long by human standards. A fraction of a second is quite sufficient. (See Fig. 2-13.)

Bounce-free switches will not be necessary for every manually operated switch in every digital circuit. But when the need arises, this simple project can make the difference between a functional, high-technology piece of equipment and a useless piece of junk.

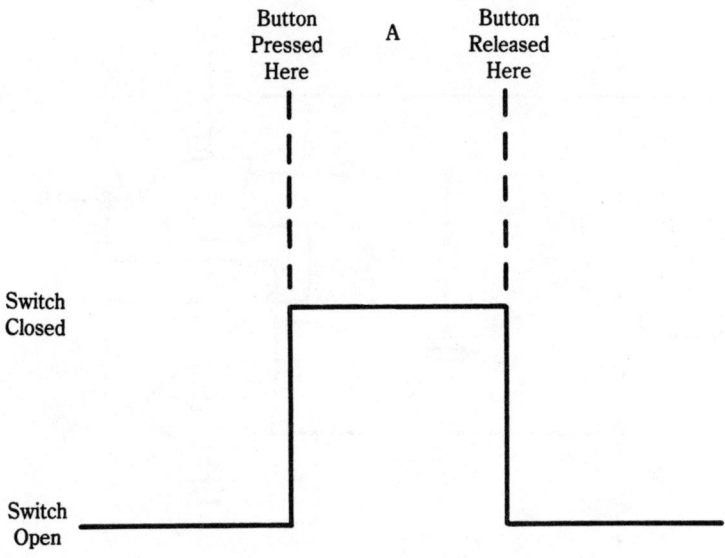
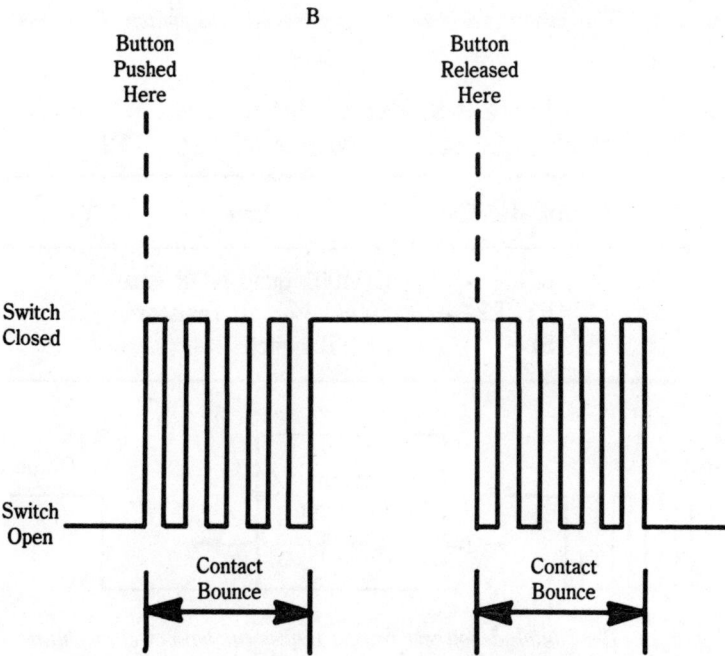

Fig. 2-11. Real mechanical switches tend to "bounce."

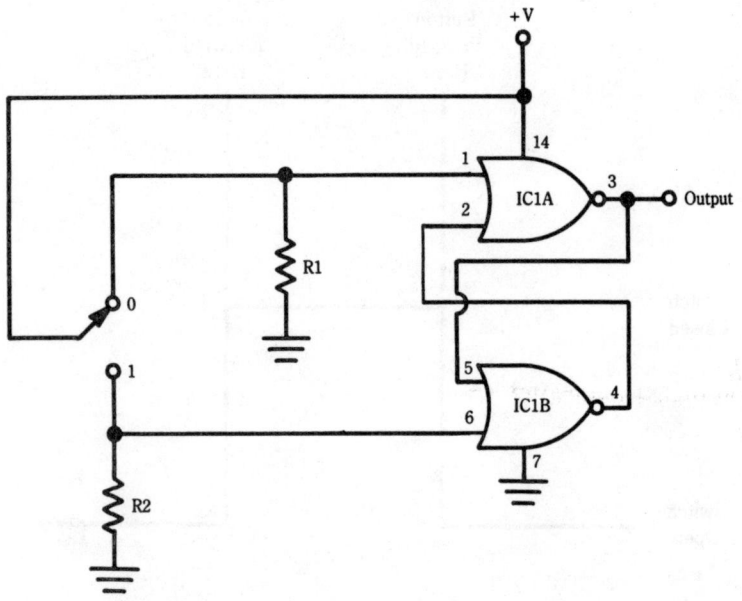

Fig. 2-12. This circuit "debounces" any mechanical switch. Project 8

Table 2-8. Parts List for the Switch Debouncer Project of Fig. 2-12.

Schematic Label	Part
IC1	CD4001 quad NOR gate
R1, R2	100K, ¼-watt resistor
S1	SPDT switch

Fig. 2-13. The switch debouncer project makes the bounciest mechanical switch function like an ideal switch.

PROJECT 9: PROGRAMMABLE TIMER

Monostable multivibrator circuits, or timers, are useful, but not terribly exciting projects. Usually built around the popular 555 timer IC, the output is normally Low until the circuit is triggered. Then the output goes High for a preset period of time, determined by resistance and capacitance values in the circuit, regardless of the length of the trigger pulse. When the circuit times out, the output goes Low again. Some timers have a normally High output which goes Low only during the timing cycle, but it is basically the same thing.

By using the XR2240 in place of the 555 we can create a more versatile, and far more interesting timer. This circuit, shown in Fig. 2-14 is programmable. The length of the timing cycle can be varied simply by flicking a few switches. No component values have to be changed. What's even more exciting is that the circuit is programmable over a very wide range. The ratio between the shortest to longest time is 1:255. A single circuit can take care of almost all your timer needs. You could almost call this circuit a "universal timer".

The XR2240 has eight outputs. Each operates at a specific multiple of the base time (T). The output values increase in binary fashion:

PIN No.	MULTIPLE
1	1T
2	2T
3	4T
4	8T
5	16T
6	32T
7	64T
8	128T

Output values can be combined to create any timing value from 1T to 255T. If two or more output pins are used together, simply by connecting them simultaneously to the circuit output, the timing values can be added together. For example, let's say the following pans are connected to the circuit

35

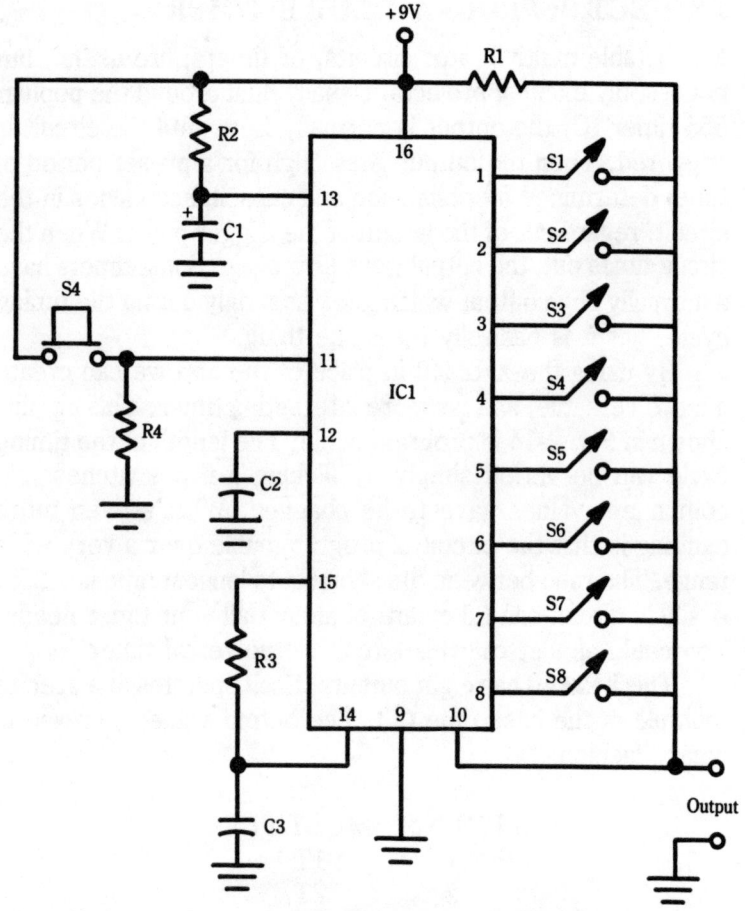

Fig. 2-14. This timer circuit is programmable. Project 9

output—1, 4, and 6. In this case, the output's timing value will be:

$$1T + 8T + 32T = 41T$$

If all eight output pins are used simultaneously, the total timing value will be at the maximum possible value:

$$1T + 2T + 4T + 8T + 16T + 32T + 64T + 128T = 255T$$

The base time is determined by the values of resistor R2 and capacitor C1. The formula couldn't possibly be simpler:

$$T = R2C1$$

where R2 is expressed in ohms and C1 is expressed in farads.

R2's value should be between 1K (1000 ohms) and 10 Meg (10,000,000 ohms). The recommended range for C1 is from $0.01\mu F$ (0.00000001 farad) to $1000\mu F$ (0.001 farad). Thus the base time can be set at anything from 10 μS (0.00001 second) up to 10,000 seconds, or 2 hours, 46 minutes, 40 seconds. If the time base is set at its maximum and all eight outputs are used (255T), the timing period will last 708 hours, 20 minutes. Quite an impressive range!

To design for a specific time base, just select a likely value for C1, and rearrange the time base equation to solve for the value or R2:

$$R2 = T/C1$$

In this project, we will use a time base of 1 second. If we use a 10 μF capacitor, then R2 should have a value of:

$$\begin{aligned}R2 &= 1/0.00001 \\ &= 100{,}000 \text{ ohms} \\ &= 100K\end{aligned}$$

The complete parts list for this project is given in Table 2-9. Note that if you want to use a different time base, only R2 and C1 need to be changed.

With a time base of 1 second, a time period from 1 to 255 seconds (4¼ minutes) can be set via switches S1–S8, in one second intervals. This project can be used as a versatile and precise kitchen or darkroom timer, among many other possible applications.

Pushbutton switch S9 and resistor R4 are used to manually trigger the beginning of the timing cycle. If you choose to use some external signal to trigger the circuit, simply eliminate

Table 2-9. Parts List for the Programmable Timer Project of Fig. 2-14.

Schematic Label	Part
IC1	XR2240 programmable timer
C1	10 μF 15V electrolytic capacitor*
C2, C3	0.01 μF capacitor
R1	10K, ¼-watt resistor
R2	100K, ¼-watt resistor*
R3	22K, ¼-watt resistor
R4	1 Megohm, ¼-watt resistor**
S1–S8	SPST switch
S9	Normally Open SPST push button**

*Frequency determining component
**Optional component

these two components and feed the trigger signal directly into pin 11.

Depending on your specific application, you may use separate switches for S1–S8, or you can use a pair of four unit DIP switches. A DIP switch is the size of an eight pin IC, except it has four tiny switches on its back. These switches may be set with a pencil or a small screwdriver. DIP switches are used for switches that aren't changed frequently, especially when space is at a premium. A four unit DIP switch package fits in the space that a single average sized switch would take up.

PROJECT 10: 24-HOUR TIMER

Common timer circuits can have timing periods ranging from a fraction of a second up to a few hours. That is quite a wide range, and covers a great many applications; but a few applications require longer timing periods.

Fig. 2-15 shows an astable multivibrator circuit built around the XR2240 programmable timer IC. Using the component values listed in Table 2-10 will give a total cycle time of 24 hours.

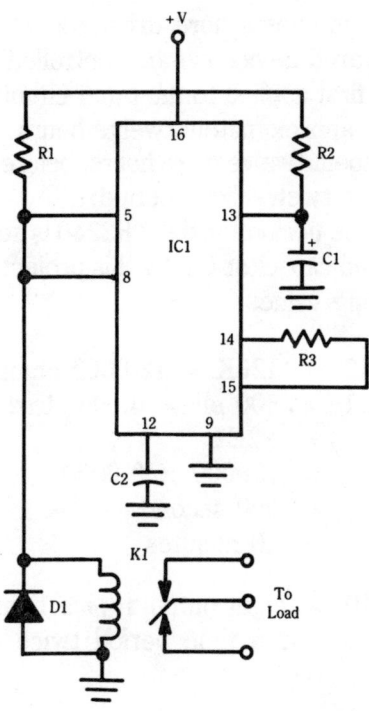

Fig. 2-15. This timer circuit has a 24 hour timing cycle.   Project 10

Table 2-10. Parts List for the 24 Hour Timer Project of Fig. 2-15.

Schematic Label	Part
IC1	XR2240 programmable timer
D1	1N4002 diode
K1	relay with contacts rated to suit desired load
C1	500 μF 35V electrolytic capacitor*
C2	0.01 μF capacitor
R1	10K, ¼-watt resistor
R2	120K, ¼-watt resistor
R3	22K, ¼-watt resistor

*timing components

The output is shown here driving a relay. Almost any electrically powered device can be controlled by this circuit. When power is first applied to the timer circuit, the relay will be activated for approximately twelve hours. Then it will be deactivated for about twelve more hours, before being activated again for another twelve hour period.

The base time period for the XR2240 is set by the values of resistor R2 and capacitor C1. In this project we are dealing with the following values:

$$R2 = 120K = 120,000 \text{ ohms}$$
$$C1 = 500 \ \mu F = 0.005 \text{ farad}$$
$$T = R2C1$$
$$= 120000 \times 0.005$$
$$= 600 \text{ seconds}$$
$$= 10 \text{ minutes}$$

The XR2240 has eight outputs (pins 1 through 8). Each successive output has a time period twice as long as the preceding one:

PIN	MULTIPLE	TIME PERIOD
1	1T	10 minutes
2	2T	20 minutes
3	4T	40 minutes
4	8T	80 minutes
5	16T	160 minutes
6	32T	320 minutes
7	64T	640 minutes
8	128T	1280 minutes

Twenty-four hours equals:

$$60 \times 24 = 1440 \text{ minutes}$$

so pins 5 and 8 can be combined for a total of approximately twenty four hours:

$$160 + 1280 = 1440 \text{ minutes}$$

The on and off times for this circuit won't be quite as neat and precise as an actual clock, but for many applications it will be close enough. For example, if the application is to turn the lights on at night and off in the morning, who really cares if the lights are switched on at 6:57, or 7:02, instead of 7:00 at the dot. For such a "quick and dirty" circuit, this project does a remarkably good job.

PROJECT 11: LONG-DURATION TIMER

Timer ICs, like the popular 555, offer a wide range of timing periods, ranging from a fraction of a second, up to several minutes. This is more than sufficient for the vast majority of practical applications. But once in a while, you will need a circuit with a longer timing period.

The XR2240 programmable timer, discussed in earlier projects, is capable of extremely long timing periods. But let's assume this chip is not available. There is another approach to extending the timing period of a monostable multivibrator.

Practically speaking, the upper limit for the 555 is about ten or fifteen minutes, unless a special (and expensive) low-leakage capacitor is used for the timing capacitor. Even if you can afford such a low-leakage capacitor, it will probably be difficult to find one with the desired value.

It is usually more practical to achieve longer time periods by cascading two or more timer stages, as illustrated in Fig. 2-16. But to keep this project within limitations of the title of this book, you can use a 556 dual timer, or a 558 quad timer.

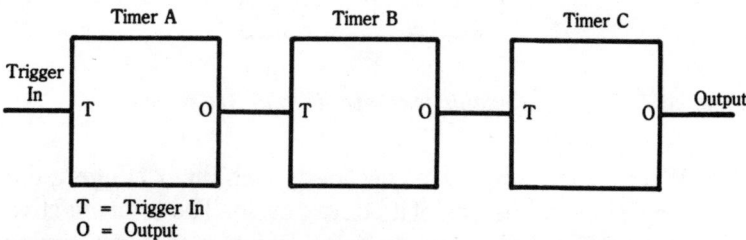

Fig. 2-16. Longer timing periods can be achieved by cascading two or more timer stages.

The 558 is particularly well suited for cascading timer stages for long duration time periods. The pin-out diagram for this chip is shown in Fig. 2-17. The 558 IC contains four separate timers that are similar to the 555. To reduce the number of pins, the timers in the 558 are somewhat simplified. This chip is designed for use in monostable applications only. It can't be used in astable multivibrator circuits.

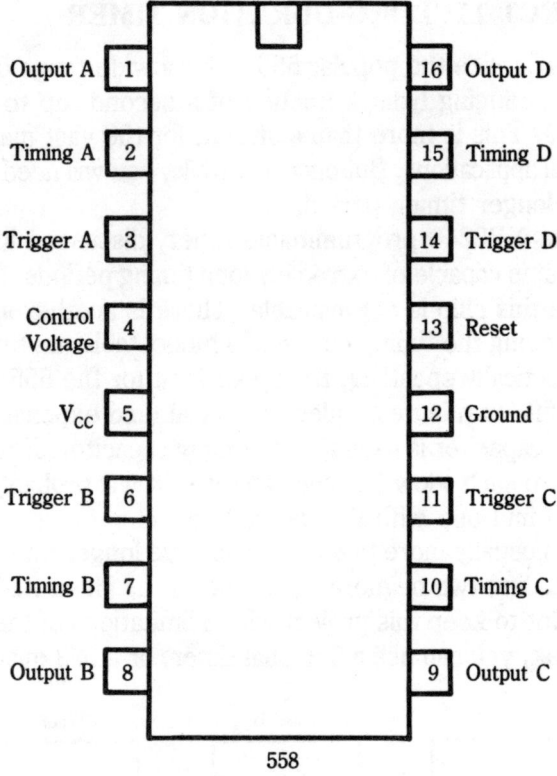

Fig. 2-17. The 558 contains four separate 555-type timers.

When timer stages are cascaded, each timer triggers the next one in line at the end of its timing cycle. The total effective timing period is found simply by adding together the timing periods of each timer stage.

The formula for the 558's timing period is the same as the equation used for the 555:

$$T = 1.1RC$$

A four-stage, cascaded long-duration timer project is shown in Fig. 2-18.

Since there are four timer stages, there are four separate time constants to set up:

STAGE 1 TIME	=	$1.1R_1C_1$
STAGE 2 TIME	=	$1.1R_3C_2$
STAGE 3 TIME	=	$1.1R_5C_3$
STAGE 4 TIME	=	$1.1R_7C_4$

The LED and current-dropping resistor R8 provide a visual indication of the output for demonstration purposes. You can use the circuit to drive other circuits and output devices. The final output is at pin 16.

Resistors R2, R4, and R6 should have equal values. The exact resistance is not critical, as long as these three resistors are matched. Anything between 3.3K and 68K should do.

A typical parts list for this project is given in Table 2-11. You will probably want to experiment with other component values. The components marked with asterisks (*) in the parts list are timing components.

Using the values given in the parts list, the timing periods for each of the stages are as follows:

T1	=	$1.1R_1C_1$
R1	=	1 Megohm = 1,000,000 ohms
C1	=	100 μF 0.0001 farad
T1	=	1.1 × 1000000 × 0.0001
	=	110 seconds
T2	=	$1.1R_3C_2$
R3	=	820K = 820,000 ohms
C2	=	250 μF = 0.00025 farad
T2	=	1.1 × 820000 × 0.00025

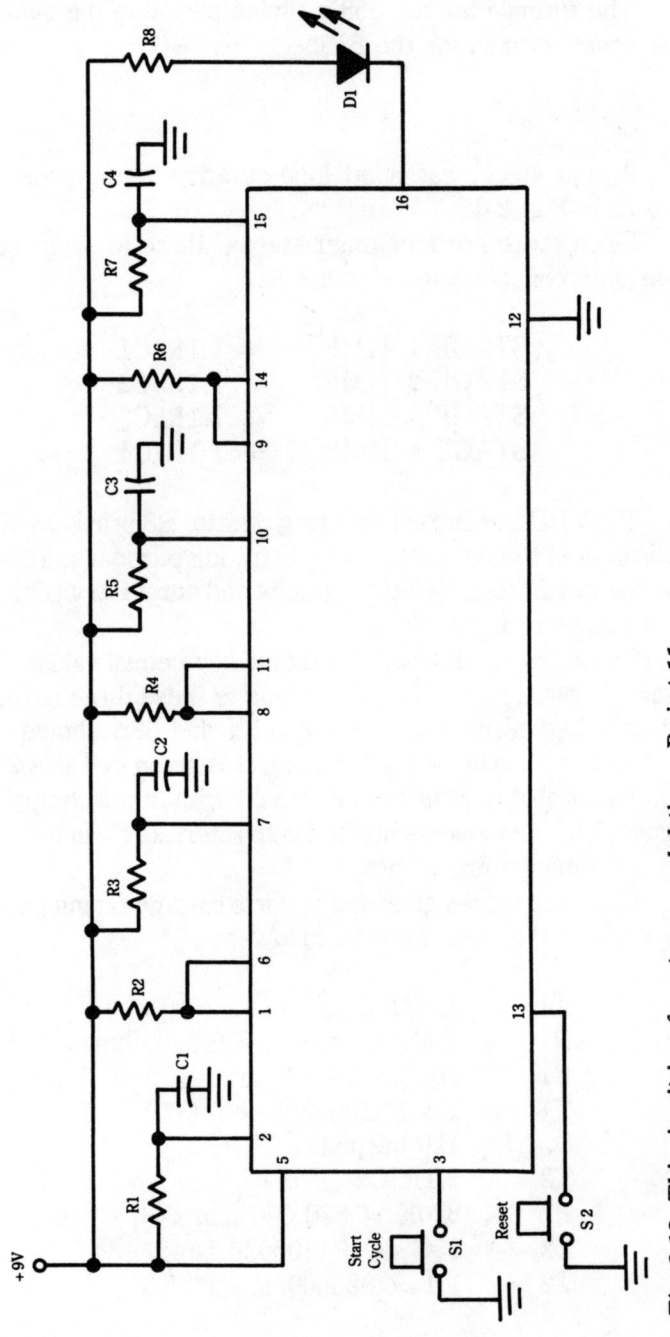

Fig. 2-18. This circuit is a four stage cascaded timer. Project 11

44

Table 2-11. Parts List for the Long Duration Timer Circuit of Fig. 2-18.

Schematic Label	Part
IC1	558 quad timer
D1	LED
R1	1 Megohm, ¼-watt resistor
R2, R4, R6	4.7K, ¼-watt resistor
R3	820K, ¼-watt resistor
R5	680K, ¼-watt resistor
R7	1.5 Megohm, ¼-watt resistor
C1	100 µF, 35 Volt electrolytic capacitor
C2	250 µF, 35 Volt electrolytic capacitor
C3, C4	470 µF, 35 Volt electrolytic capacitor
S1, S2	Normally Open SPST push switch

$$
\begin{aligned}
&= 225.5 \text{ seconds} \\
T_3 &= 1.1 R_5 C_3 \\
R_5 &= 680K = 680{,}000 \text{ ohms} \\
C_3 &= 470 \,\mu F = 0.00047 \text{ farad} \\
T_3 &= 1.1 \times 680000 \times 0.00047 \\
&= 351.56 \text{ second} \\
T_4 &= 1.1 R_7 C_4 \\
R_7 &= 1.5 \text{ Megohm} = 1{,}500{,}000 \text{ ohms} \\
C_4 &= 470 \,\mu F = 0.00047 \text{ farad} \\
T_4 &= 1.1 \times 1500000 \times 0.00047 \\
&= 775.5 \text{ seconds}
\end{aligned}
$$

The total effective timing period is simply the sum of the individual stage times:

$$
\begin{aligned}
T_t &= T_1 + T_2 + T_3 + T_4 \\
&= 110 + 225.5 + 351.56 + 775.5 \\
&= 1{,}462.56 \text{ seconds} \\
&= 24 \text{ minutes, } 22.56 \text{ seconds}
\end{aligned}
$$

Of course, other component values will result in other timing periods.

PROJECT 12: TIMED TOUCH SWITCH

This is a sort of combination project. It combines a timer with a touch switch, like the one described in Project No. 1. Actually the timer in this case is a switch debouncer circuit. The schematic for this project is shown in Fig. 2-19, with the parts list appearing in Table 2-12.

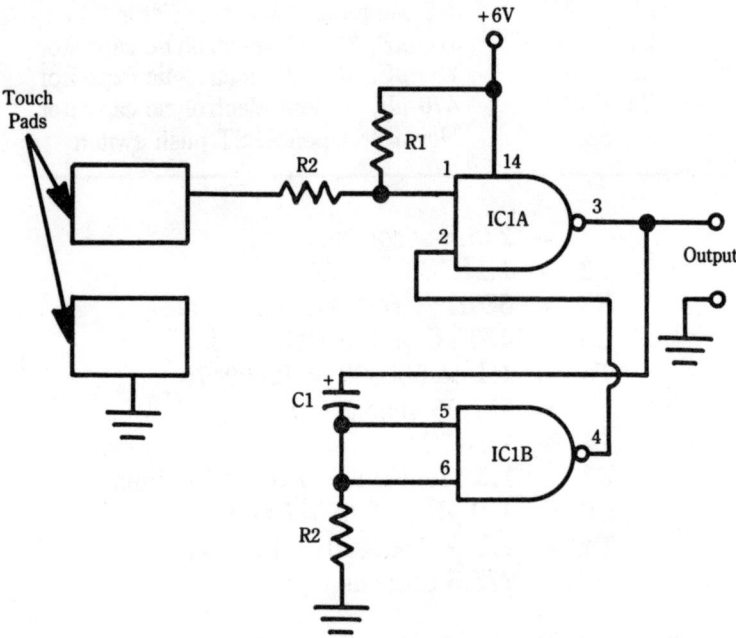

Fig. 2-19. This is a timed touch switch circuit. Project 12

When the circuit is activated by shorting the two touch plates with your finger, as shown in Fig. 2-20, the output will go High for a specific period of time, and then go Low again, even if the touch plates are being shorted continuously. For the component values given in the parts list, the time period

Table 2-12. Parts List for the Timed Touch Switch Project of Fig. 2-19.

Schematic Label	Part
IC1	CD4011 quad NAND gate
C1	5 µF 15 Volt electrolytic capacitor
R1	10 Megohm, ¼-watt resistor
R2, R3	120K, ¼-watt resistor

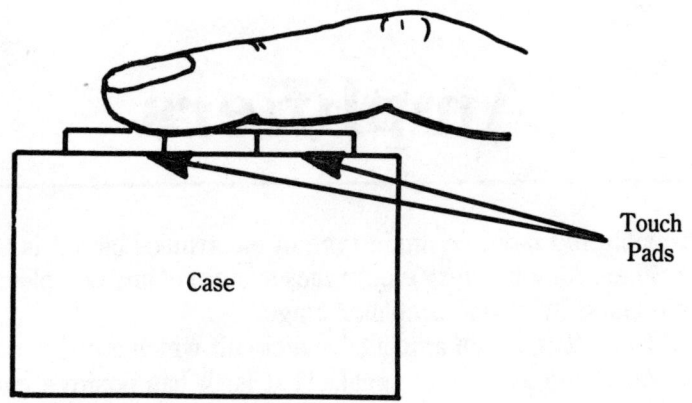

Fig. 2-20. The timed touch switch is activated by shorting the two touch plates with your finger.

will be approximately one second. Experiment with other component values.

As with any touch switch, this project should be powered from batteries *only*. Refer back to the discussion of Project No. 1.

3
Amplifiers

Probably the most common type of electronics circuit is the amplifier. Almost every electronics system of any complexity includes at least one amplifier stage.

By definition, an amplifier is a circuit which amplifies, or increases the level of a signal. That is, it has positive gain. The input signal multiplied by the gain is the output signal. There should be no difference between the input and output signals except for amplitude. Any other change in the waveshape would be considered distortion.

Some amplifier circuits do not have positive gain. Maybe they shouldn't really be called amplifiers, but since the circuitry is essentially the same, what else are we going to call them?

Many amplifier circuits have unity gain. In other words, the gain is exactly one. The output signal is identical in level to the input signal. This might seem like the most pointless circuit imaginable, but it is really very useful in many applications. A unity gain amplifier is called a buffer. It is used for impedance matching between stages, or to minimize loading problems.

A few amplifier circuits actually have negative gain. The output signal is at a lower level than the input signal. These circuits might be called attenuators.

This chapter features amplifiers and related projects for a number of applications.

PROJECT NO. 13: AUDIO AMPLIFIER

Audio amplifiers are always useful projects. They amplify signals within the audible range of approximately 20 Hz to 20 kHz.

This project is designed around the LM380 audio amplifier IC. This chip has been around for a few years now, and it is still quite popular. It is inexpensive, reasonably powerful, and easy to use.

The LM380 is widely available in two types of package styles. The pin-out for the compact eight pin DIP version is shown in Fig. 3-1. The fourteen pin DIP version is illustrated in Fig. 3-2. On both versions of the LM380 only six of the pins are active. The remaining pins should be shorted to ground so that they will provide limited internal heat-sinking. Obviously, the fourteen pin version has more internal heat-sinking than the eight pin version. This means that the larger version can handle greater amounts of power without overheating.

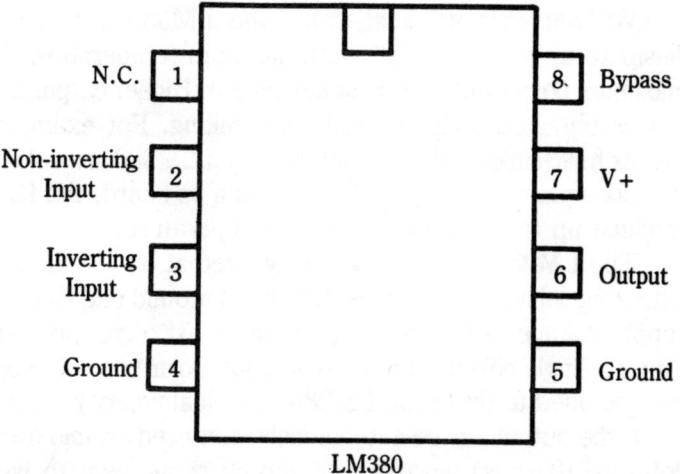

Fig. 3-1. The LM380 audio amplifier IC is available in an eight pin DIP housing.

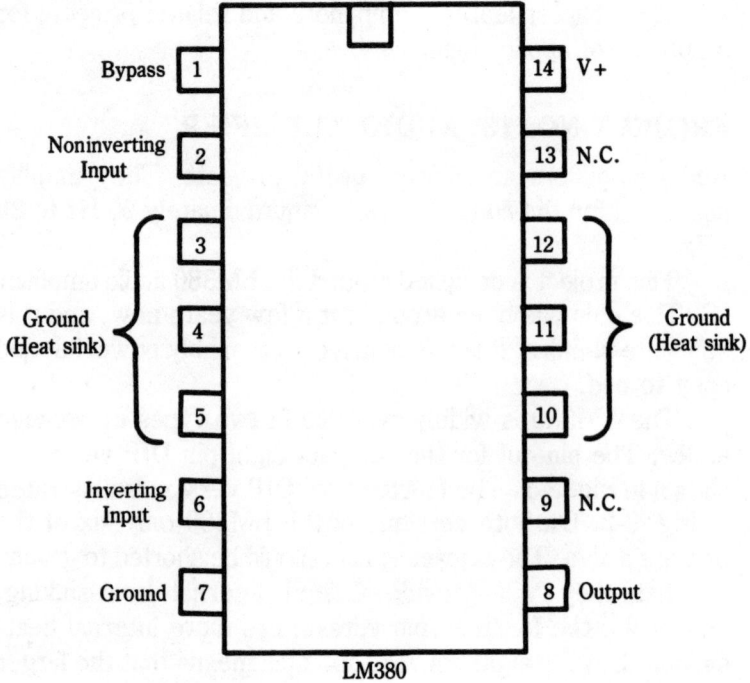

Fig. 3-2. The LM380 is also available in a fourteen pin version.

Without external heat-sinks, the LM380 can typically dissipate up to about 1.25 watts at room temperature. The maximum power output is heat dependent. Higher output levels can be obtained with external heat-sinking. For example, if the six heat-sinking pins of a fourteen pin LM380 are soldered to a six square inch copper foil pad on a pc board, the IC can produce up to 3.7 watts at room temperature.

The LM380's gain is internally fixed at 50 (34 dB). The output signal automatically centers itself around one half of the supply voltage, effectively eliminating most offset problems. Either a single polarity (positive) or a dual polarity power supply may be used to drive the LM380. If a dual polarity supply is used, the output will be automatically centered around ground potential (0 volts) with virtually no dc component to worry about. An output capacitor is still advisable to compensate for any inequality between the two halves of the voltage supply.

The input stage of the LM380 is rather unique. The input signal may be either referenced to ground, or ac coupled, depending on the specific requirements of the individual application.

The inputs are internally biased with a 150K resistance to ground. Transducers, or previous stages, which are referenced to ground (no dc component in the signal) may be directly coupled to either the inverting or the non-inverting input. In most applications, only one of the two inputs is used. There are several possible ways to handle the unused input terminal. They are:

- leave it floating (unconnected)
- short it directly to ground

or:

- reference it to ground through a resistor or a capacitor.

The LM380 is designed for use with a minimum of external components, which certainly makes life easier for the circuit designer. The most basic form of the LM380 circuit is shown in Fig. 3-3. Clearly, it would be hard for things to be much simpler than this. The only external component is the output decoupling capacitor, which may be optional in some applications.

Notice that the fourteen pin version of the LM380 is shown in this diagram. The larger version is recommended for this project. If you decide to use the eight pin version, correct the pin numbering (refer to Figs. 3-1 and 3-2). It would be a particularly good idea to use an external heat-sink with the eight pin version. External heat-sinking would not be a bad idea for the fourteen pin version either, but it is not as critical with the larger chip.

Of those grounded pins in Fig. 3-3, pin 7 is the supply ground connection point. The remaining grounded pins, 3, 4, 5, 10, 11, and 12, are the internal heat-sinking pins.

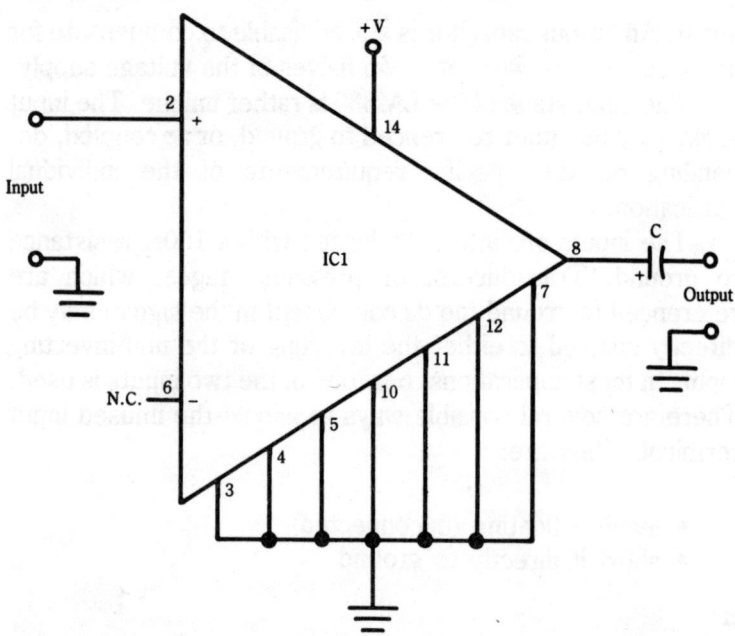

Fig. 3-3. The LM380 is designed to be used with minimum external components.

While this super-simple circuit is functional in many practical applications, several additional external components may be required to obtain the best performance or to add specific features.

If the LM380 chip is physically located more than two or three inches from the power supply filter capacitor, adding an extra decoupling capacitor is advisable. This capacitor should be mounted between the V+ terminal of the LM380 and ground. A supply line decoupling capacitor should always be mounted as physically close to the body of the IC as possible. A typical value for this component is 0.1 µF.

Now we're ready for our audio amplifier project. The schematic appears in Fig. 3-4. The parts list is given in Table 3-1.

Notice that the heat-sinking pins are not shown in the schematic diagram. This omission is intended to reduce diagram clutter. The heat-sinking pins are always grounded in all applications, so the connections can be assumed.

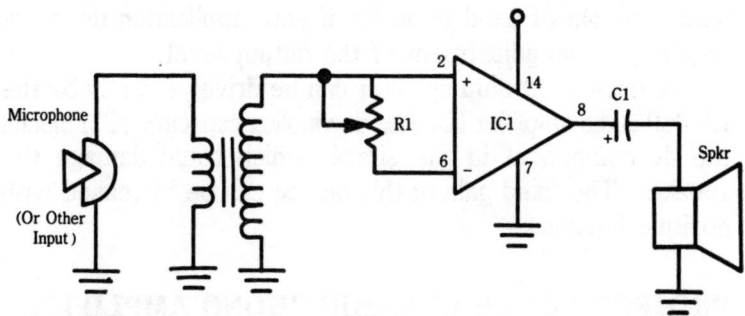

Fig. 3-4. This is a practical audio amplifier circuit using the LM380. Project 13

Table 3-1. Parts List for the Audio Amplifier Circuit of Fig. 3-4.

Schematic Label	Part
IC1	LM380 audio amplifier
T1	impedance matching transformer PRIMARY: 500 ohms (or to match input) SECONDARY: 200K
C1	500 μF, 35 Volt electrolytic capacitor
R1	1 Megohm potentiometer (audio taper)
SPKR	small 8 ohm speaker

The input to this circuit is provided by an inexpensive low-impedance microphone. Other low level signal sources may be used in place of this microphone if you choose. The impedance matching transformer (T1) should be used with any low-impedance source. If the signal source has a high output impedance, you may omit this transformer.

The input signal should not have too high an amplitude, or the amplifier may be overloaded. This could result in severe distortion in the output signal.

The one megohm potentiometer (R1) serves as a volume control for the amplifier. You could replace the potentiometer with a couple of fixed resistors if your application does not require manual adjustment of the output level.

A small eight ohm speaker can be driven directly by the LM380 audio amplifier IC. The decoupling capacitor (C1) blocks the dc component in the signal, which could damage the speaker. The fixed gain of this device can be increased with positive feedback.

PROJECT NO. 14: CERAMIC PHONO AMPLIFIER

Do you have an old record player lying around that doesn't work? If the turntable turns, but the amplifier is dead, then this is just the project for you. It is a replacement amplifier for a ceramic type cartridge. The schematic diagram for this project is shown in Fig. 3-5.

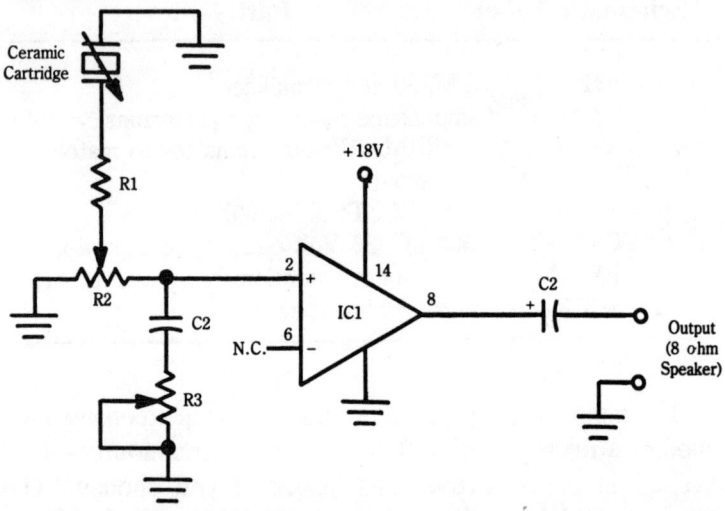

Fig. 3-5. This amplifier is designed to be used with a ceramic cartridge phonograph. Project 14

There really isn't much to be said about this project. It has two potentiometer controls. Potentiometer R1 is the volume control, while potentiometer R3 is a simple tone control. With R3, the user can tailor the sound somewhat to suit individual tastes. This control adjusts the high frequency roll-off characteristics of the circuit.

This project is built around the LM380 audio amplifier IC. This chip is an inexpensive and easy to use low power amplifier that can be used in a great many audio applications.

The parts list for this project is given in Table 3-2.

**Table 3-2. Parts List for
the Ceramic Phono Amplifier Circuit of Fig. 3-5.**

Schematic Label	Part
IC1	LM380 audio amplifier
C1	0.05 µF capacitor
C2	500 µF 35 volt electrolytic capacitor
R1	75K, ¼-watt 5% resistor
R2	25K potentiometer
R3	10K potentiometer

PROJECT NO. 15: AUDIO MIXER

Fig. 3-6 shows a simple but useful audio mixer preamplifier. The parts list is given in Table 3-3.

Three independent inputs are shown here and you can easily expand the circuit for additional inputs. Input A passes through C1, R1, and R2. Capacitor C1 blocks any dc component in the signal. Potentiometer R1 is a gain control for that particular input. The other input circuits are similar.

Potentiometer R8 is a master gain control, affecting all the inputs equally. Capacitor C5 prevents any dc component or offsets in the output signal from reaching the next stage in the system—probably a power amplifier.

For experimental or non-critical applications, the 741 op amp IC will do just fine. But for serious audio applications, you should substitute a high-grade low-noise op amp IC.

PROJECT NO. 16: SIGNAL SPLITTER

A signal splitter is essentially a mixer in reverse. A mixer combines two or more isolated inputs into a single output. A signal splitter takes a single input signal and feeds it out to two or more isolated outputs.

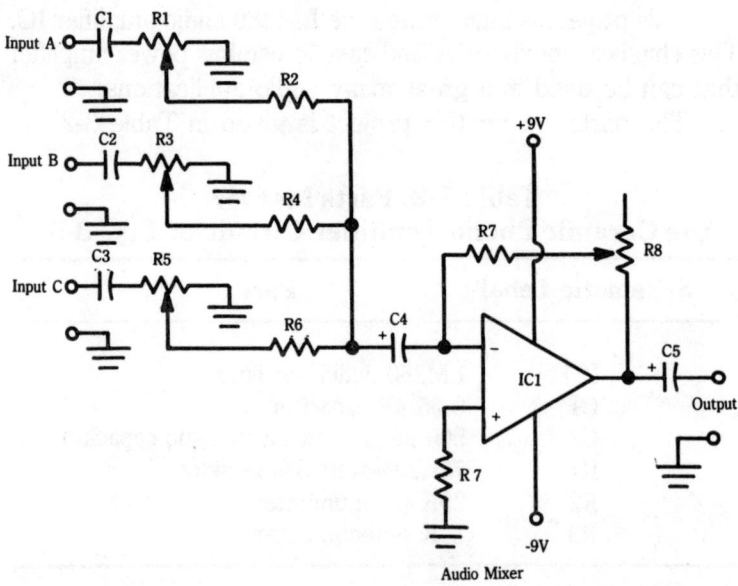

Fig. 3-6. An audio mixer is used to combine multiple input signals into a single output. Project 15

Table 3-3. Parts List for the Audio Mixer Project of Fig. 3-6.

Schematic Label	Part
IC1	op amp (741, or similar—see text)
C1-C3	0.5 µF capacitor
C4	1 µF, 15V electrolytic capacitor
C5	2.5 µF 15V electrolytic capacitor
R1, R3, R5	50K potentiometer (audio taper)
R2, R4, R6	120K, ¼-watt resistor
R7	100K, ¼-watt resistor
R8	500K potentiometer (audio taper)

A practical signal splitter circuit is shown in Fig. 3-7. The parts list is given in Table 3-4.

The amplitude of each output is individually controllable via potentiometers R5, R7, and R9. The capacitors block any dc component in the signal. While three output channels are

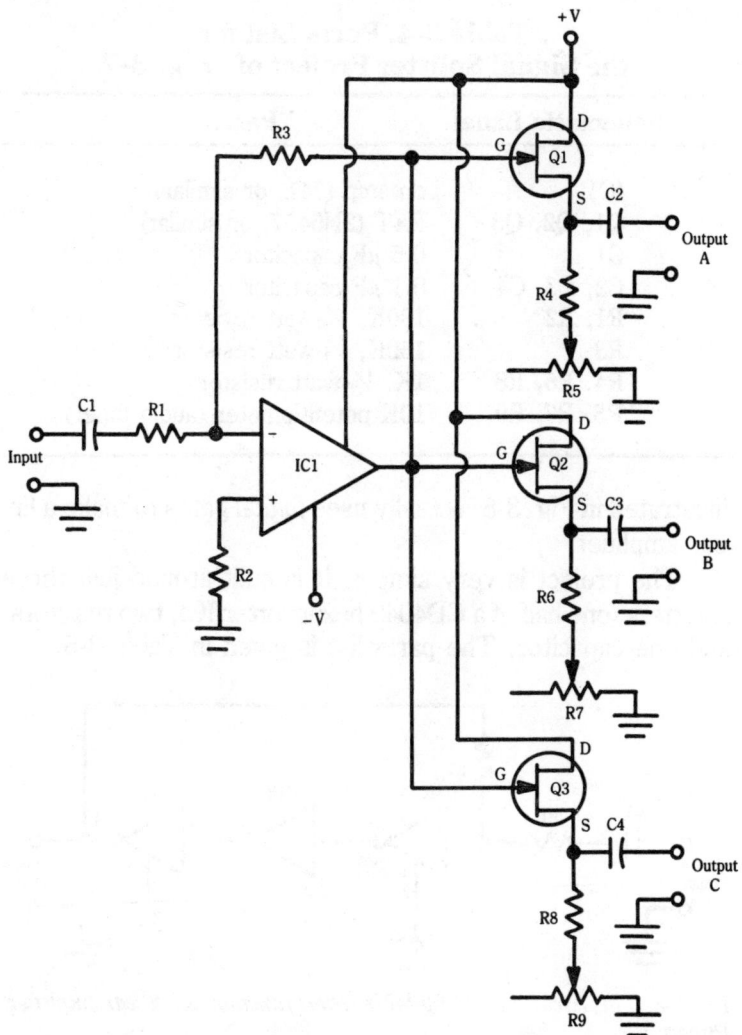

Fig. 3-7. A signal splitter takes a single input signal and feeds it out to multiple independent outputs. Project 16

shown here, you can easily expand the project to permit additional output channels.

PROJECT NO. 17: DIGITAL LINEAR AMPLIFIER

This is one of my favorite projects because it is so simple and so unusual. In fact, it is almost weird. This circuit, which is

Table 3-4. Parts List for the Signal Splitter Project of Fig. 3-7.

Schematic Label	Part
IC1	op amp (741, or similar)
Q1, Q2, Q3	FET (2N5457, or similar)
C1	0.5 µF capacitor
C2, C3, C4	0.1 µF capacitor
R1, R2	100K, ¼-watt resistor
R3	150K, ¼-watt resistor
R4, R6, R8	1K, ¼-watt resistor
R5, R7, R9	10K potentiometer (audio taper)

illustrated in Fig. 3-8, actually uses digital gates to make a linear amplifier.

The project is very simple. It is built around just three inverters (one half of a CD4049 hex inverter IC), two resistors, and one capacitor. The parts list is given in Table 3-5.

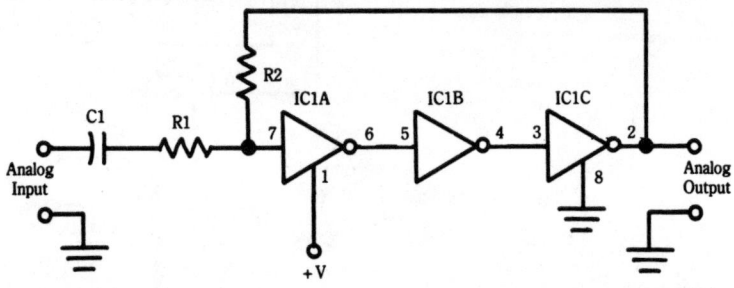

Fig. 3-8. This circuit makes digital inverters function as a linear amplifier. Project 17

The gain of this amplifier circuit is determined by the ratio of the values of resistors R2 and R1:

$$G = R2/R1$$

You are encouraged to experiment with various resistor values. Of course, if R1 is less than R2, the circuit will exhibit

Table 3-5. Parts List for the Digital Linear Amplifier Project of Fig. 3-8.

Schematic Label	Part
IC1	CD4049 hex inverter
C1	0.01 μF capacitor
R1	100K, ¼-watt resistor
R2	1 Megohm, ¼-watt resistor

negative gain (attenuation). For the component values given in the parts list, the gain is 10. The purpose of the capacitor is to block any dc component in the input signal. Its value is not critical.

PROJECT NO. 18: DIFFERENCE AMPLIFIER

The op amp (or operational amplifier) was originally designed to perform mathematical operations. It isn't always apparent, but in most of its applications, the op amp is performing a mathematical operation of some sort. For example, the basic inverting and non-inverting amplifier circuits are performing simple multiplication. That is the input multiplied by the circuit gain equals the output.

The op amp's most fundamental mathematical operation is subtraction. This function is performed by the difference, or differential, amplifier. The basic circuit, with closed loop feedback to control the circuit gain is illustrated in Fig. 3-9.

Generally, the two feedback loops, inverting and non-inverting will be given equal component values. That is:

$$R1 = R2$$
$$R3 = R4$$

This exact component equality is not absolutely necessary, although it is usually convenient. However, the feedback ratios must be equal for the circuit to function properly. That is:

$$\frac{R3}{R1} = \frac{R4}{R2}$$

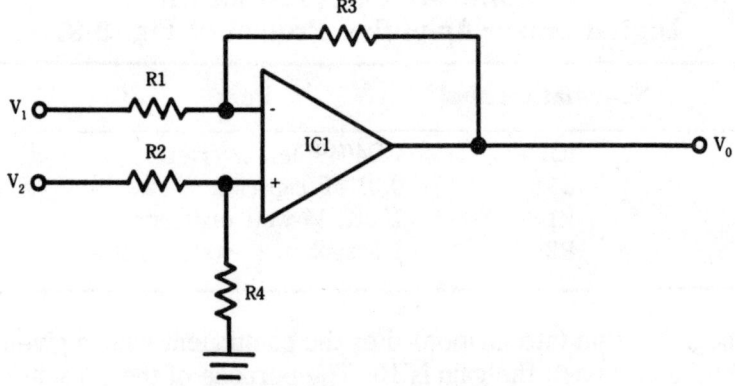

Fig. 3-9. A difference amplifier performs subtraction.

These ratios define the closed loop gain of the circuit. They need to be equal to give balanced gain for the two inputs. The gain can be determined from either feedback loop:

$$G = \frac{R3}{R1} = \frac{R4}{R2}$$

If all four resistors have the same value, the circuit will exhibit unity gain. The exact resistance value used is irrelevant. Remember, the gain is determined by the resistance ratios, not by the resistances themselves.

With unity gain, the circuit's operation as a subtractor is most obvious. If we call the signal applied to the non-inverting input Va, and the voltage fed to the inverting input Vb, then the output will be equal to:

$$Vo = Va - Vb$$

A demonstration circuit for a difference amplifier is shown in Fig. 3-10. A typical parts list is given in Table 3-6. Experiment with other values for the gain determining resistors (R1–R4).

Resistors R5 through R8 are voltage dividers to permit you to feed manually controllable dc voltages to the two inputs. To use the project with external signals, eliminate these

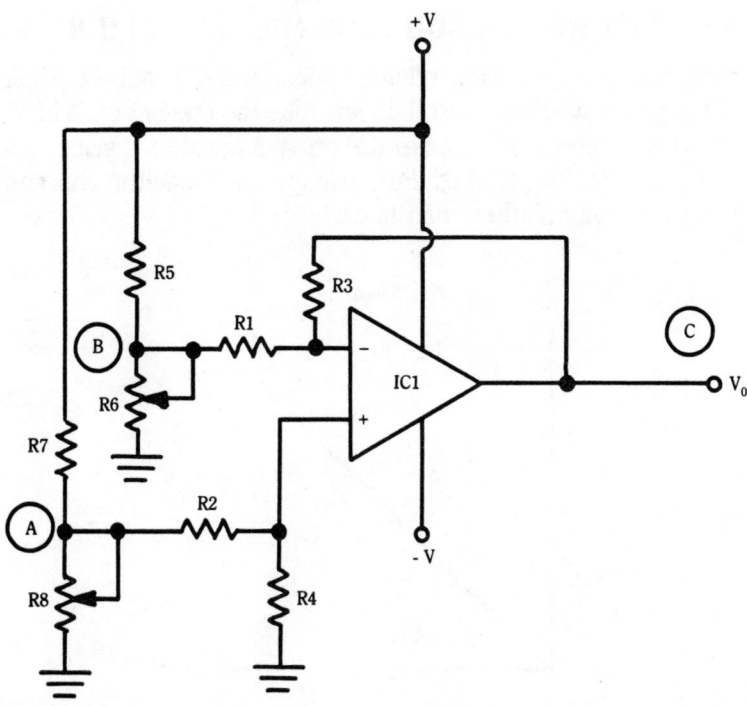

Fig. 3-10. This circuit demonstrates the operation of a difference amplifier. Project 18

Table 3-6. Parts List for the Difference Amplifier Project of Fig. 3-10.

Schematic Label	Part
IC1	op amp (741, or similar)
R1–R4	10K, ¼-watt resistor
R5, R7	1K, ¼-watt resistor
R6, R8	1K potentiometer

resistors. While the difference amplifier is most frequently used with dc signals, there's no reason why you can't try feeding in some ac inputs. Some very peculiar waveforms can result from subtracting two out-of-phase (and probably different frequency) ac signals.

61

PROJECT NO. 19: LOGARITHMIC AMPLIFIER

Most analog applications will use a linear scale. That is, a graph of the action would be a straight line, like the one in Fig. 3-11A. But some phenomena operate on a logarithmic scale, as illustrated in Fig. 3-11B. For example, a capacitor charges logarithmically, rather than linearly.

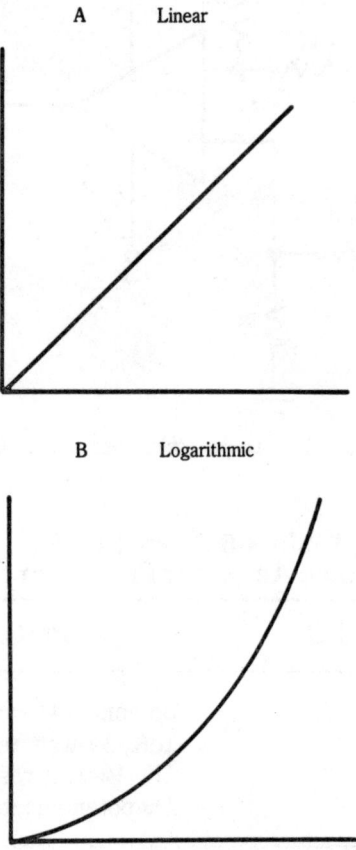

Fig. 3-11. Some functions are linear (A), while others are logarithmic (B).

The logarithm of a number is the exponent that indicates the power to which another "base" number must be raised to produce the given number. "Natural" logarithms use a base of 10. Here are a few numbers and their natural logarithms:

$$\text{Log } 1 = 0$$
$$\text{Log } 2 = 0.30103$$
$$\text{Log } 3 = 0.04771213$$
$$\text{Log } 4 = 0.60206$$
$$\text{Log } 5 = 0.69897$$
$$\text{Log } 10 = 1.0$$
$$\text{Log } 12 = 1.0791812$$
$$\text{Log } 20 = 1.30103$$
$$\text{Log } 25 = 1.39794$$
$$\text{Log } 30 = 1.4771213$$
$$\text{Log } 100 = 2.0$$
$$\text{Log } 1000 = 3.0$$
$$\text{Log } 10000 = 4.0$$
$$\text{Log } 100000 = 5.0$$

Raising 10 to the logarithm produces the original number:

$$10^0 = 1\infty$$
$$10^{0.30103} = 2$$
$$10^{0.4771213} = 3$$
$$10^1 = 10$$
$$10^2 = 100$$
$$10^3 = 1000$$

Raising any number to the power of zero (X^0) always results in 1, so the logarithm of 1 is always 0, no matter what base is used. The logarithm of zero is infinity in the negative direction:

$$\text{Log } 0 = \infty$$

Negative logarithms indicate numbers less than one:

$$\text{Log } 0.1 = -1$$
$$\text{Log } 0.2 = -0.69897$$
$$\text{Log } 0.005 = -2.30103$$

To work from the logarithm back to the original number is called finding the antilogarithm. For example, the antilog of

0.30103 is 2, the antilog of 1 is 10, and so forth. The base does not necessarily have to be ten, but we won't concern ourselves with other bases here.

By placing a transistor in the feedback path of an op amp, in place of the ordinary resistor, we can make a logarithmic amplifier. The basic circuit is shown in Fig. 3-12. A linear change in the input signal will produce a logarithmic change in the output signal.

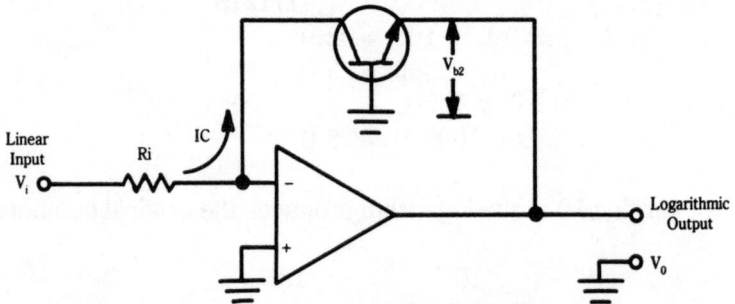

Fig. 3-12. Placing a transistor in the feedback path of an op amp creates a logarithmic amplifier.

As you can see in the diagram, the logarithmic amplifier is basically an inverting amplifier circuit with a reverse-biased transistor as the feedback component.

Logarithmic amplifiers are used to perform multiplication and other mathematical operations. It might be interesting to experiment with feeding audio signals through a logarithmic amplifier.

The secret of this circuit is the fact that many semiconductor pn junctions, diodes and transistors, exhibit pretty good logarithmic characteristics.

The op amp's output is applied across the base–emitter junction of the transistor. In other words:

$$V_0 = V_{b2}$$

When the reverse bias across the transistor is large enough, typically -100 mV, or more, the resulting collector current leads to an output voltage with a value of:

$$V_0 = V_{b2}$$
$$= -2.3 \times q \times \text{Log}[I_C/(z \times I_{esat})]$$

where q is a temperature sensitive constant. It is typically about 0.026 at room temperature, simplifying the equation to:

$$V_0 = V_{b2}$$
$$= -0.0598 \times \text{Log}[I_c/(z \times I_{esat})]$$

where z is the common base current gain of the transistor, and I_{esat} is the emitter saturation current. Both of these values can be obtained from the manufacturer's specification sheet for the transistor. I_C, of course, is the collector current.

The collector current is equal to the current flowing through the input resistor, R_i. Therefore, the above equation can be re-written as:

$$V_0 = V_{b2}$$
$$= -0.06 \times \text{Log}[(V_i/R_i)/(z \times I_{esat})]$$

Notice that we have rounded off the constant factor here.

Without going into the rather complex mathematics, the formula can be further reduced to:

$$V_0 = V_{b2}$$
$$= -0.06 \times \text{Log}(V_i) - V_{0.05}$$

$V_{0.05}$ is a small output offset voltage. Its value depends on the value of R_i, and the temperature of the transistor. This offset voltage is typically very small, so in many applications it can be ignored, leaving:

$$V_0 = V_{b2}$$
$$= -0.06 \times \text{Log}(V_i)$$

External components may also be used to compensate for the output offset voltage to bring it closer to true zero. You can see that the output is proportional to the logarithm of the input value.

By reversing the positions of the resistor and the transistor, as shown in Fig. 3-13, we get an antilogarithmic amplifier. This works in the opposite manner as the logarithmic amplifier.

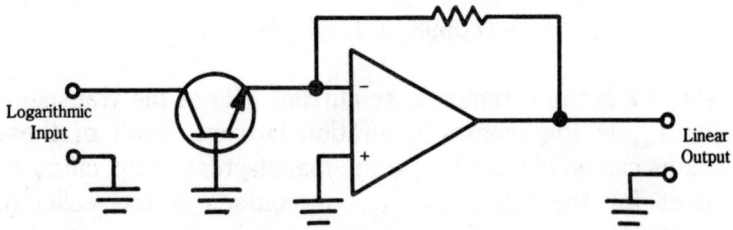

Fig. 3-13. An antilogarithmic amplifier reverses the action of a logarithmic amplifier.

Moving on from the theory to the actual project, Fig. 3-14 shows a practical logarithmic amplifier circuit. The parts list is given in Table 3-7.

Capacitor C1 and diode D1 help smooth out the output. The diode also blocks negative feedback signals. R4 is a trimpot, is used to calibrate the circuit, compensating for any offset signals within the IC. With an input voltage of exactly 1.0 volt, carefully adjust R4 for an output voltage of exactly 0.0 volt. Remember, the logarithm of 1 is always 0. This is all there is to calibrating the logarithmic amplifier. Now, for any voltage signal at the input, the circuit will produce the logarithmic value at the output, unless the limits of the IC are exceeded. It will be very difficult to drive the op amp into saturation in this application, because the input voltage increases at a much faster rate than the output voltage. The op amp will be damaged by excessive input signals long before it can be driven into saturation.

For the greatest accuracy, monitor the output of the logarithmic amplifier with a good digital voltmeter. An analog voltmeter will be difficult to read precisely.

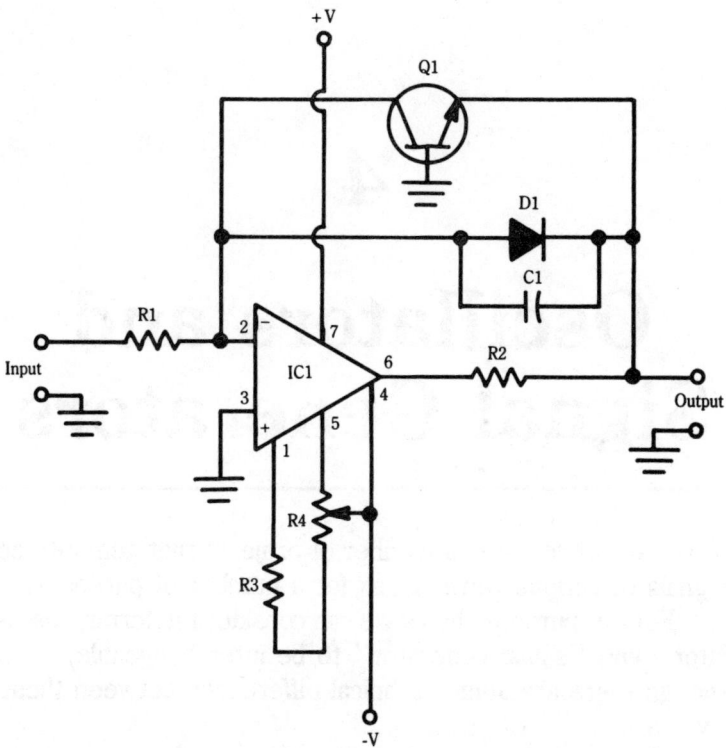

Fig. 3-14. This is a practical logarithmic amplifier circuit. Project 19

Table 3-7. Parts List for the Logarithmic Amplifier Circuit of Fig. 3-14.

Schematic Label	Part
IC1	op amp (741, or similar)
Q1	NPN transistor (2N2222, or similar)
D1	1N4001 diode
C1	0.1 µF capacitor
R1	47K resistor
R2	2.2K resistor
R3	330 ohm resistor
R4	10K trimpot

4

Oscillators and Signal Generators

This chapter features a number of projects that generate ac signals of various waveshapes for a number of purposes.

For our purposes here, we can consider the terms "oscillator" and "signal generator" to be interchangeable, even though there are some technical differences between these two terms.

The circuits used in these projects range from some very basic, commonly used circuits to some very unusual circuits. All of these projects can easily be modified for a variety of applications.

PROJECT NO. 20:
OP AMP SQUARE-WAVE GENERATOR

The circuit shown in Fig. 4-1 generates clean square waves with a precise 1:2 duty cycle. This means the output is High for exactly one half of each complete cycle. The square wave includes all odd harmonics, and no even harmonics.

This is a very easy circuit to work with, because only a handful of resistors and a single capacitor are required as external components.

The frequency of the square wave output signal is determined by the values of C1, R3, and R4, and the gain of the

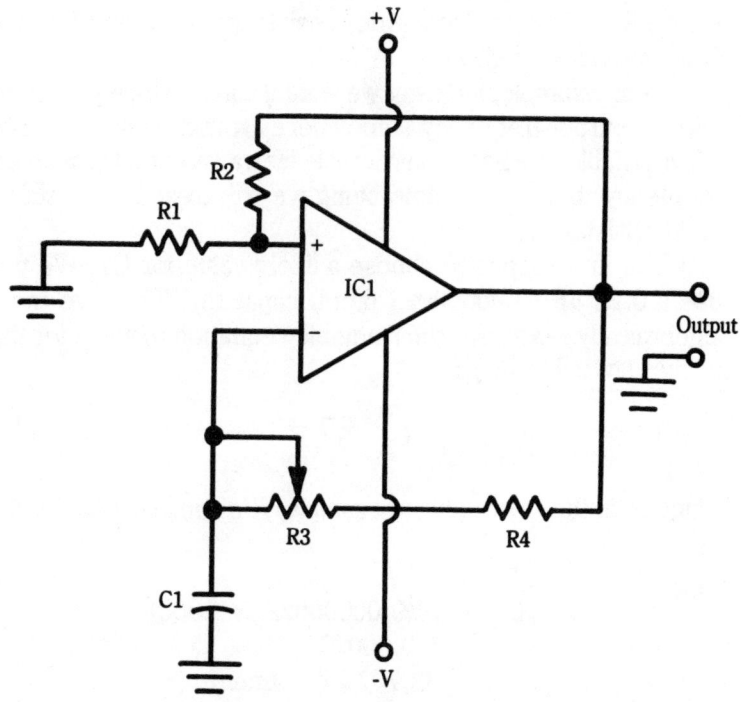

Fig. 4-1. This is a simple op amp square-wave generator circuit. Project 20

op amp. Larger component values will result in lower frequencies. R3 is a manually variable potentiometer, allowing the user to adjust the output frequency while the circuit is in operation. If only a single fixed-output frequency is required, R3 and R4 can be combined into a single fixed resistor.

Resistors R1 and R2 control the overall gain of the op amp and set the output level of the square wave signal. The output frequency is influenced by the amplifier gain, so it is a good idea to stick to a gain of 10. That is, the value of R2 should be ten times larger than the value of R1. If the gain is 10, then the output frequency can be determined with this simple formula:

$$F = 5/RC$$

where R is the combined series resistance of R3 and R4, and C is the value of C1.

As an example, let's say we want a square wave generator with an output frequency somewhere around 1000 Hz. Since R3 is variable, we have considerable leeway around the nominal frequency, but for the time being we will treat R as a single fixed resistance.

The first step is to choose a likely value for C1. We will use a 0.22 µF (0.00000022 farad) capacitor. Then we must algebraically rearrange the frequency equation to solve for the necessary value of R:

$$R = 5/CF$$

Plugging in the values for our example, R's nominal value works out to:

$$\begin{aligned} R &= 5/(0.00000022 \times 1000) \\ &= 5/0.00022 \\ &= 22{,}727.272 \text{ ohms} \end{aligned}$$

If we wanted a fixed frequency square wave generator, we'd simply use a 22K resistor as R, and not bother with R3. For a variable frequency circuit, we will need to do a little more figuring, but at this point there is no need at all for precision. Rough approximations will do, because the final circuit will be tunable.

The relative size of R3's maximum value compared to R4's fixed value will determine the range of the circuit. If R3's maximum value is small, while R4 is relatively large, the square wave generator will have a fairly narrow range. If, on the other hand, R3 has a rather large maximum value, and R4's resistance is low, the circuit will have a wide tuning range. Remember, however, that the wider the tuning range, the harder it will be to precisely adjust the potentiometer for a specific output frequency. A more narrow range will give you greater precision in fine-tuning the circuit.

For most applications, the best compromise will probably be to give R3 and R4 more or less equal values at the design frequency, 1000 Hz in our example. The design frequency should be at the center of the potentiometer's range, so R3's maximum value should be about twice as large as R4's fixed value.

In our example, the combined series resistance of R3 and R4 should be approximately 22K for an output frequency of 1000 Hz. Splitting this in half, R3 and R4 should each have a value of 11K. This is not a standard resistance value, so let's use a 10K resistor for R4. At 1000 Hz, R3 should be about 12K. If this was the potentiometer's mid-point, the maximum resistance should be 24K. Again, this is not a common value, but a 25K potentiometer will be very close to the theoretical ideal.

Now, let's consider how this circuit works. To begin, we will assume that the op amp's output is in positive saturation. That is, the output signal is just slightly below V+. The non-inverting input is held at a specific fraction of the positive saturation voltage by the voltage divider made up of R1 and R2.

Meanwhile, the capacitor is being charged through R3 and R4. The capacitor's voltage is fed into the op amp's inverting input. At some point, the charge on the capacitor will exceed the voltage at the non-inverting input. Since the inverting input signal is now greater than the non-inverting signal, it takes over and drives the output negative. The op amp quickly goes into negative saturation (just above V−).

The voltage divider presents the same fraction of the output voltage back to the non-inverting input, except now the polarity is reversed. The voltage is negative, rather than positive. The capacitor now charges negatively through R3 and R4 until it again exceeds the voltage at the non-inverting input. The output is driven positive, and the entire cycle repeats itself.

Since the capacitor obviously has the same charging rate for both polarities, because the same components are used for both halves of the cycle, the positive saturation time will be equal to the negative saturation time. This means that the output signal will be a neat, symmetrical square wave with a 1:2, or 50 percent, duty cycle.

A parts list for a 1000 Hz square wave generator is given in Table 4-1.

Table 4-1. Parts List for the Op-amp Square-wave Generator Project of Fig. 4-1.

Schematic Label	Part
IC1	op amp (741, or similar)
C1	0.22 µF capacitor
R1, R4	10K, ¼-watt resistor
R2	100K, ¼-watt resistor
R3	25K potentiometer

For other duty cycles, diodes may be used to create two separate feedback paths as shown in Fig. 4-2. R3 and R4 control the time the output is positive, and R5 and R6 control how long it is negative. A limitation of this modified circuit is that changing the duty cycle will alter the output frequency.

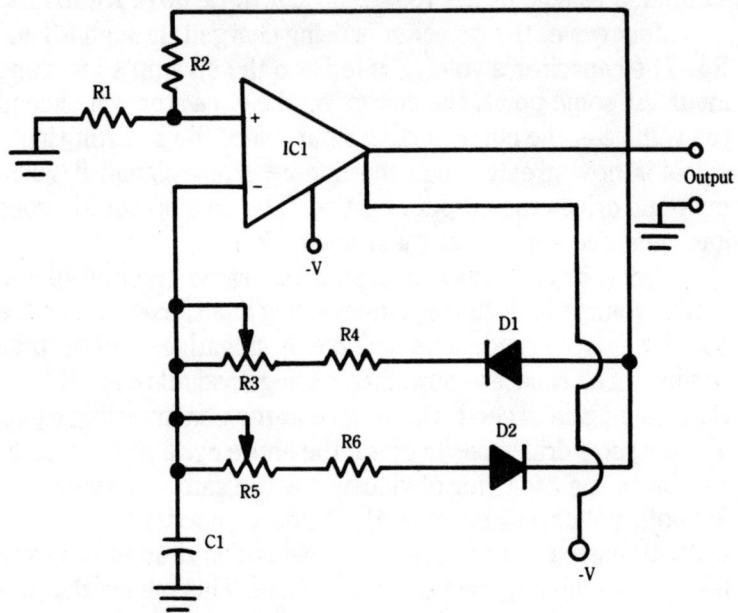

Fig. 4-2. A simple modification can be made to the circuit of Fig. 4-1 to generate rectangular waves with duty cycles other than 50 percent.

PROJECT NO. 21: PROGRAMMABLE SQUARE-WAVE GENERATOR

Figure 4-3 shows a programmable square wave generator circuit. This project is built around the XR2240 programmable timer IC, which was discussed in Project 9.

In this circuit, the XR2240 is used in its astable mode. The base frequency for this circuit is determined by the values of resistor R2 and capacitor C1, according to this simple formula:

$$F = 1/(2R2C1)$$

The various output values act as frequency dividers. In other words, the output frequency will be equal to the base frequency divided by the timing value:

$$F_o = F/n$$

where n is the timing value of the selected outputs from 1 to 255. Obviously, the base frequency, available at output pin 1-1T, will always be the highest available frequency.

The component values listed in Table 4-2 are for a base frequency of approximately 15 kHz (15,000 Hz). If you wish to use a different base frequency, only the values of R2 and C1 will have to be changed. All other components keep the same values, regardless of the frequency.

Starting with a 0.01 μF (0.00000001 farad) capacitor as C1, the necessary value of R2 for a base frequency of 15,000 Hz works out to:

$$\begin{aligned} R2 &= 1/(2FC1) \\ &= 1/(2 \times 15000 \times 0.00000001) \\ &= 1/0.0003 \\ &= 3333.3333 \text{ ohms} \\ &= 3.3K \text{ resistor} \end{aligned}$$

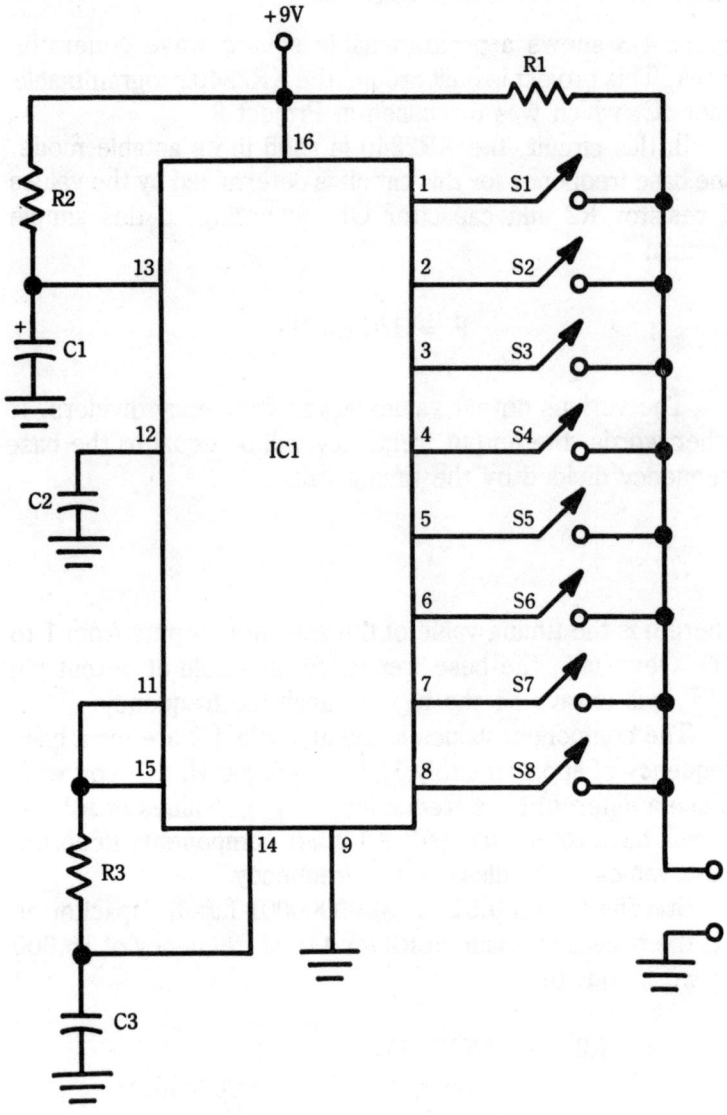

Fig. 4-3. This square-wave generator circuit is programmable. Project 21

Table 4-2. Parts List for the Programmable Square-wave Generator Project of Fig. 4-3.

Schematic Label	Part
IC1	XR2240 programmable timer
C1	0.01 µF capacitor*
C2, C3	0.01 µF capacitor
R1	10K, ¼-watt resistor
R2	3.3K, ¼-watt resistor*
R3	22K, ¼-watt resistor
S1–S8	SPST switches (may use DIP switches)

*Frequency determining component

The output frequency (F_o) for each of the eight output pins is:

PIN NO.	VALUE	Fo
1	1	15000 Hz
2	2	7500 Hz
3	4	3750 Hz
4	8	1875 Hz
5	16	937.5 Hz
6	32	469 Hz
7	64	234 Hz
8	128	117 Hz

The frequency values are approximate. There will be some variation due to component tolerances.

Intermediate timing values can be obtained by combining two or more output lines (closing more than one of the eight output switches, S1-S8). For example, if we close switches S2, S4, and S5 at the same time, the timing value will be:

$$Tt = 2 + 8 + 16$$
$$= 26T$$

So the output frequency with this combination is:

$$F_o = F/T$$
$$= 15000/26$$
$$= 577 \text{ Hz}$$

With the component values in the parts list, the output frequency can range from a high of 15000 Hz down to a low of 59 Hz when all eight output switches are closed—255T, certainly a very impressive range from such a simple circuit.

PROJECT NO. 22:
555 TRIANGULAR-WAVE GENERATOR

Normally, the 555 timer IC produces only rectangular waves. That is what it was designed for. However, if we "cheat" a little, we can get the 555 to generate a pretty good triangular wave.

The circuit for this project is shown in Fig. 4-4. This is one of the simplest possible triangular wave generator circuits. Aside from the 555 timer IC itself, only six external components are required. R1, R2, and C1 control the output frequency. Potentiometer R1 permits you to manually change the frequency while the circuit is in operation.

A typical parts list for this project appears in Table 4-3.

PROJECT NO. 23:
OP AMP SINE-WAVE OSCILLATOR

The simplest of all waveforms is the sine wave, illustrated in Fig. 4-5. Most waveforms consist of multiple frequency components, but the sine wave has just a single frequency component, the fundamental. Ideally there is no harmonic content at all. The sine wave is a very pure waveform.

While it is the simplest waveform, it is one of the most difficult to generate without distortion. Most practical sine wave generators put out a signal with some harmonic content. In a good design, the harmonics will be held to a very low level.

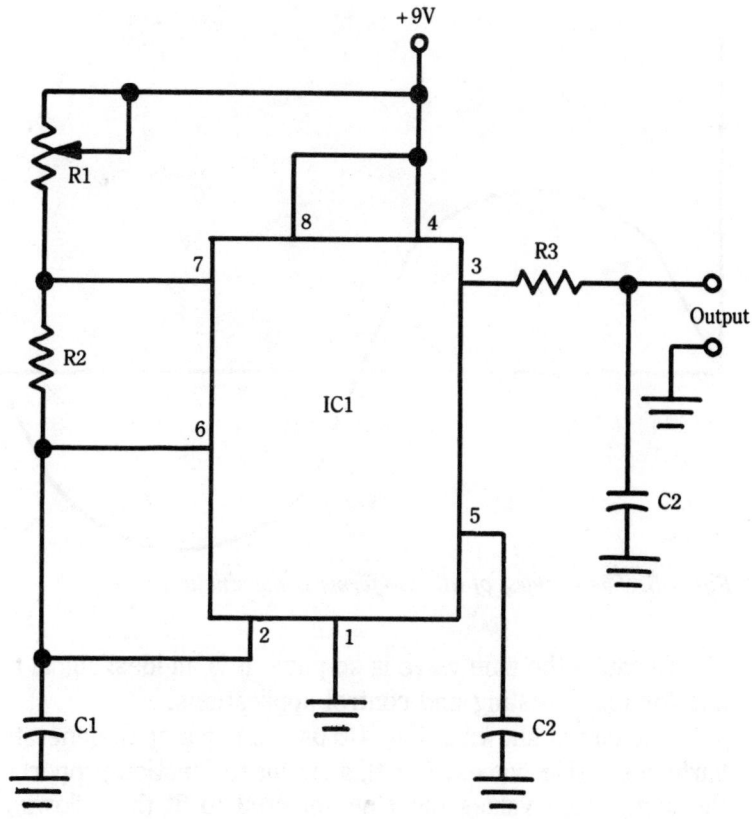

Fig. 4-4. The 555 timer IC can be used to generate triangular waves. Project 22

Table 4-3. Parts List for the 555 Triangular-wave Generator Project of Fig. 4-4.

Schematic Label	Part
IC1	555 timer
C1	0.1 µF capacitor
C2	0.022 µF capacitor
C3	0.01 µF capacitor
R1	100K potentiometer
R2	1K, ¼-watt resistor
R3	12K, ¼-watt resistor

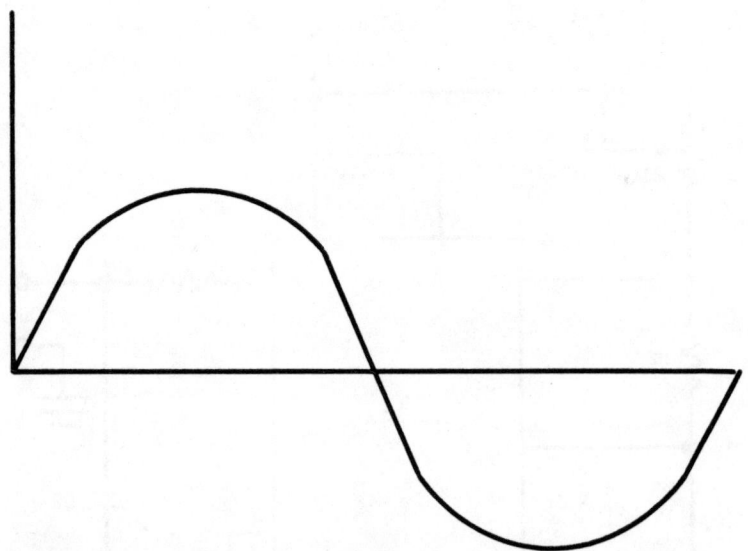

Fig. 4-5. The simplest of all waveforms is the sine wave.

Because the sine wave is so pure, it is an ideal signal to use for many testing and control applications.

The circuit shown in Fig. 4-6 uses an op amp to generate fairly clean sine waves. For this circuit to function properly, the component values must be selected to fit the following proportions:

$$
\begin{aligned}
R1 &= R3/4 \\
R2 &\cong R3/2 \\
R3 &= R4 \\
R5 &= 2R3 \\
C1 &= C2 \\
C3 &= 2C1
\end{aligned}
$$

All the passive components must be selected for the desired frequency in your particular application. The frequency of the sine wave is determined by this formula:

$$F = 1/(2\pi RC)$$

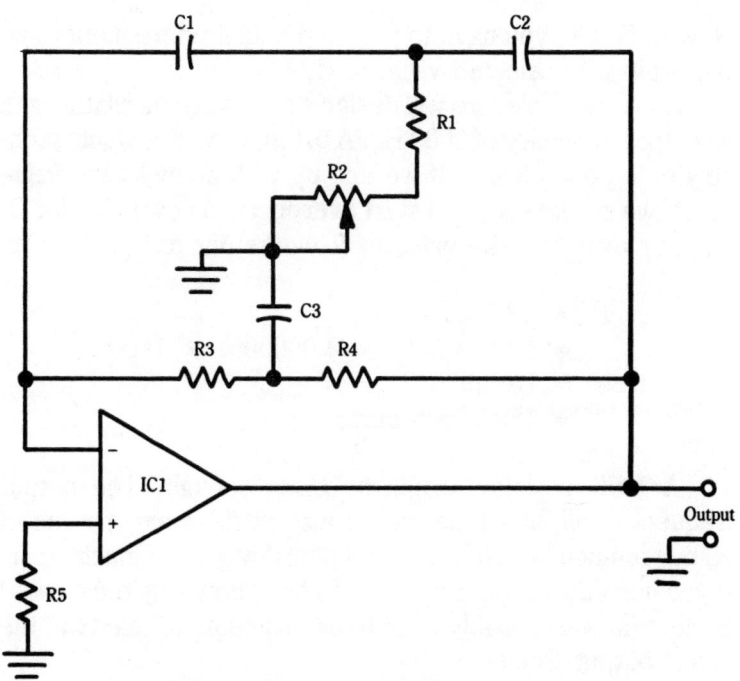

Fig. 4-6. An op amp can generate nearly pure sine waves. Project 23

where π is pi, a mathematical constant with a value of approximately 3.14, F is the output frequency in hertz, R is the value of R3 in ohms, and C is the value of C1 in farads. Since *pi* is a constant, the formula can be rewritten as:

$$F = 1/(6.28RC)$$

To determine the specific component values for a specific desired frequency, you first need to rearrange the frequency equation. Since finding a specific resistance value is usually easier than finding a specific capacitance value, your best bet will generally be to arbitrarily select a likely value for C, and rearrange the frequency equation to solve for R:

$$F = 1/(2\pi RC)$$
$$FR = 1/(2\pi C)$$
$$R = 1/(2\pi CF)$$

Now to find R, you have to plug in the desired frequency and the arbitrarily selected value of C.

As an example, we will design a sine wave oscillator with an output frequency of 1000 Hz. A 0.1 μF capacitor would probably be a good choice. If we end up with an awkward value for R, we can go back and start over, using a new value for C. For our example, the value of R works out to:

$$\begin{aligned} R &= 1/(2\pi CF) \\ &= 1/(2 \times 3.14 \times 0.0000001 \times 1000) \\ &= 1/0.000628 \\ &= 1592.3566 \text{ ohms} \end{aligned}$$

A 1.5K resistor should be close enough. The output frequency will be a little off because of the inexact value of R, but component tolerances will probably give as much error. If you need the output frequency to be at precisely the desired value, you will probably need to use trimpots to fine tune the actual output frequency.

Now it's a simple matter of assigning the appropriate value to each of the circuit components:

$$\begin{aligned} R3 &= R = 1500 \text{ ohms} \\ R1 &= R3/4 = 1500/4 = 375 = 390 \text{ ohms} \\ R2 &= R3/2 = 1500/2 = 750 \text{ ohms} \end{aligned}$$

(Use a 1K trimpot for R2.)

$$\begin{aligned} R4 &= R3 = 1500 \text{ ohms} \\ R5 &= 2R3 = 2 \times 1500 = 3000 = 3300 \text{ ohms} \\ C1 &= C = 0.1 \; \mu F \\ C2 &= C1 = 0.1 \; \mu F \end{aligned}$$

Rounding off to the nearest standard component value is entirely permissible in this circuit. The exact values aren't terribly critical, as long as the required relationships are reasonably maintained. Potentiometer R2 can be used to compensate for round off errors somewhat. Always round up on the value of R2, so that you will be able to detune the output frequency above and below its nominal value. In operation, adjust this potentiometer for the purest possible sine wave. It would be helpful to observe the waveform with an oscilloscope. The waveform should be made up of nice, smooth curves. If you don't have an oscilloscope handy, you can feed the signal through a small speaker and listen carefully as you adjust R2. Adjust the control for the purest possible tone. Generally, R2 should be a trimpot, adjusted with a screwdriver, rather than a front panel control.

This circuit puts out a reasonably pure sine wave, but only at a single frequency. To change the output frequency more than a few hertz, you will need to redesign the circuit with new component values.

The parts list for a 1000 Hz sine wave oscillator is given in Table 4-4. If you want a different output frequency, you will need to select new component values on your own. Virtually any op amp may be used. A 741 is called for in the parts list, because this is usually the cheapest and most widely available op amp IC. Of course, a higher quality op amp will generate

Table 4-4. Parts List for the
Op-amp Sine-wave Generator Project of Fig. 4-6.

Schematic Label	Part
IC1	op amp (741, or similar)
C1, C2	0.1 μF capacitor
C3	0.2 μF capacitor
R1	470 ohm, ¼-watt resistor
R2	1K potentiometer
R3, R4	1.5K, ¼-watt resistor
R5	1K, ¼-watt resistor

a somewhat purer signal. If you use a different op amp, be sure to check the pin numbering before you try to construct the project. Most op amp ICs are pin compatible with the 741, but there are some exceptions, so watch out.

PROJECT NO. 24:
DIGITAL SINE-WAVE GENERATOR

Ordinarily, digital gates work only with pulse or rectangular wave signals. It is possible, however, to generate a fair approximation of a sine wave, or almost any other analog waveform, using digital circuitry.

The basic idea is illustrated in the circuit of Fig. 4-7. IC1 is a CD4018 counter. The pin-out diagram for this chip appears in Fig. 4-8.

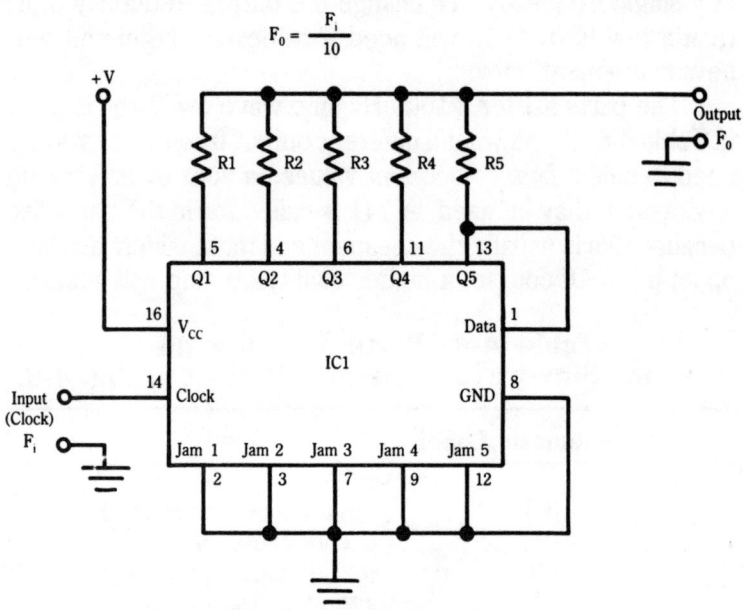

Fig. 4-7. Analog waveforms can be synthesized with a digital counter.

Each of the Q outputs is phase shifted from its predecessor by one clock pulse. The output duty cycle is always approximately 50 percent. That is, the outputs are all in the form of square waves. For odd counts, the symmetry of the output waveform will be off by plus or minus one input clock pulse. Dividing by any number from 2 to 10 is possible with the CD4018. Even larger division values can be achieved by cascading multiple CD4018s. Whatever count value is selected, the output symmetry will always be virtually the same.

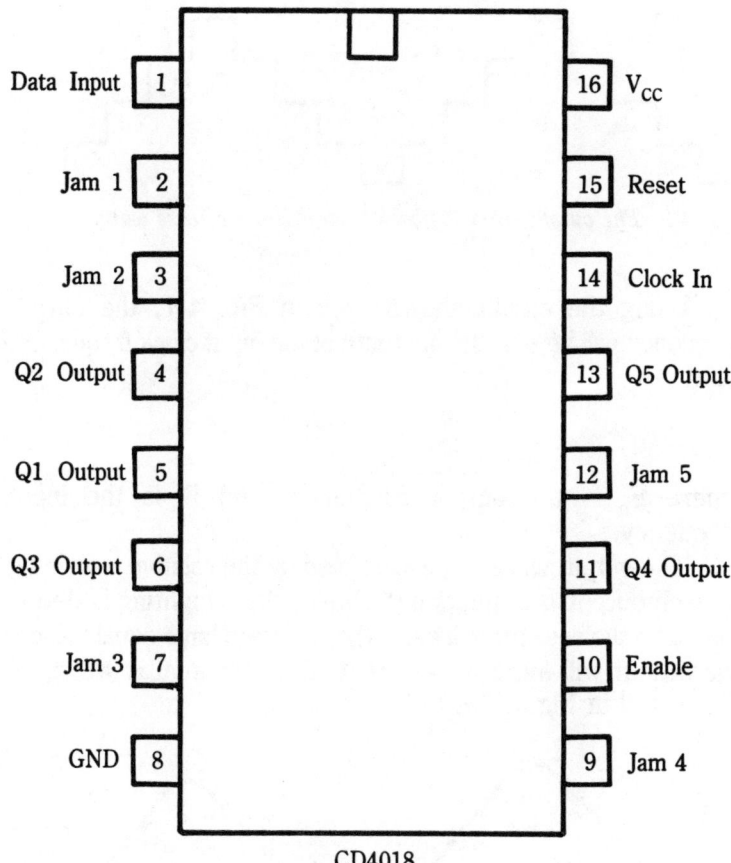

Fig. 4-8. The CD4018 counter is at the heart of this project.

The CD4018 produces several phase-shifted outputs, each one delayed by exactly one input clock pulse. If we sum the outputs together with the correct relative weighting, the result will be a staircase wave, like the one shown in Fig. 4-9. Weighting is accomplished by feeding the output signals through resistors of appropriate values.

External low-pass filtering can be added to the circuit to smooth out the steps and create a synthesized pseudo-analog waveform.

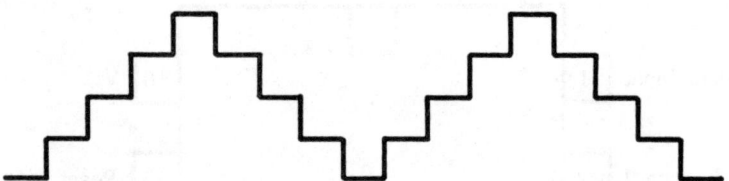

Fig. 4-9. The output from the basic circuit is a staircase wave.

Using the circuit shown back in Fig. 4-7, the output frequency will be exactly one tenth of the input clock frequency. That is:

$$F_o = F_i / 10$$

where F_o is the output frequency, and F_i is the input frequency.

The output waveshape is defined by the relative weighting of each output. As mentioned above, the weighting is determined by the resistor values. If the resistors have equal values, the output will more or less resemble a triangular wave, as illustrated in Fig. 4-10.

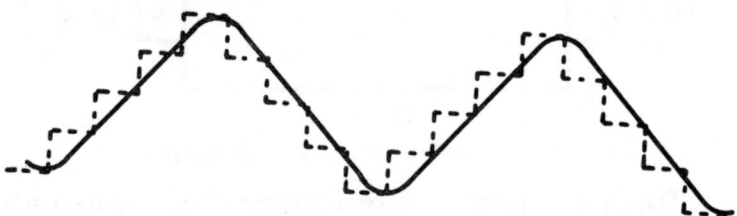

Fig. 4-10. If the resistors have equal values, the output will resemble a triangular wave.

A sine wave is similar to a triangular wave, but it has somewhat flatter peaks. The two waveforms are compared in Fig. 4-11.

We can modify the basic circuit, as shown in Fig. 4-12 to get flatter peaks. We have eliminated the output of Q5 from the waveshape even though it is still part of the counting cycle. The output frequency is still one tenth of the input clock frequency.

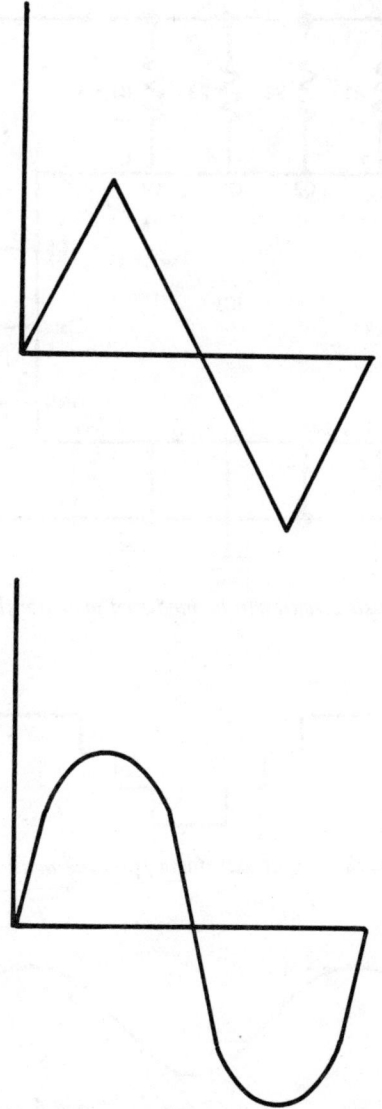

Fig. 4-11. A sine wave has flatter peaks than a triangular wave.

By eliminating Q5 from the output, the Q4 peak is effectively held for a longer portion of each cycle, as illustrated in Fig. 4-13. Filtering this signal gives us a reasonably smooth pseudo-sine wave, as shown in Fig. 4-14.

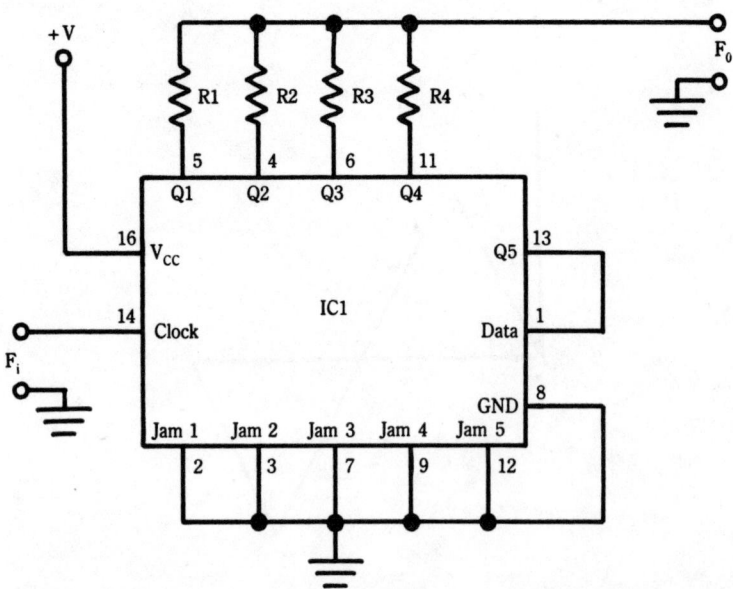

Fig. 4-12. The basic circuit can be modified to create flatter peaks in the output waveform.

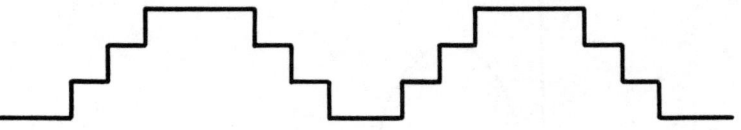

Fig. 4-13. This is the output waveform from the modified circuit of Fig. 4-12.

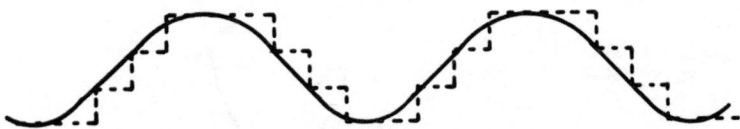

Fig. 4-14. The signal of Fig. 4-13 can be filtered to create a pseudo-sine wave.

This isn't really an ideal solution, however, since one output less is included in the waveform, reducing the resolution. That is, instead of five steps to the peak, there are only four. Still, this should be close enough for many applications. For

Table 4-6. This is the Parts List for the Digital Phaseshift Generator Project of Fig. 4-16.

Schematic Label	Part
IC1	CD4049 hex inverter
C1, C2, C3	0.01 µF capacitor
R1, R2, R3	22K, ¼-watt resistor

PROJECT NO. 26: AUDIO FUNCTION GENERATOR

A function generator is an oscillator, or signal generator circuit with multiple outputs. Two or more different waveforms may be used simultaneously. All the outputs are at the same frequency.

A circuit for an audio range function generator project is shown in Fig. 4-17. The parts list is given in Table 4-7.

This project offers three outputs:

- Rectangular wave
- Sine wave
- Triangular wave

These are probably the most commonly used waveforms in electronics testing applications.

This project is built around the 8038 function generator IC. This 14 pin chip is specifically designed for just this sort of application. The 8038 can generate sine waves, triangular waves, sawtooth waves, and rectangular waves with almost any duty cycle. The output frequency can range from a low of 0.001 Hz to over 1.0 MHz (1,000,000 Hz) with a typical distortion rating of 1.0 percent or less.

While voltage control is not used in this particular project, the 8038 can accept an external voltage to control the output frequency.

The nominal output frequency of the 8038 is set by two external timing resistors and a single external timing capacitor. The output frequency can be made manually variable by using a potentiometer for one or both of the timing resistors.

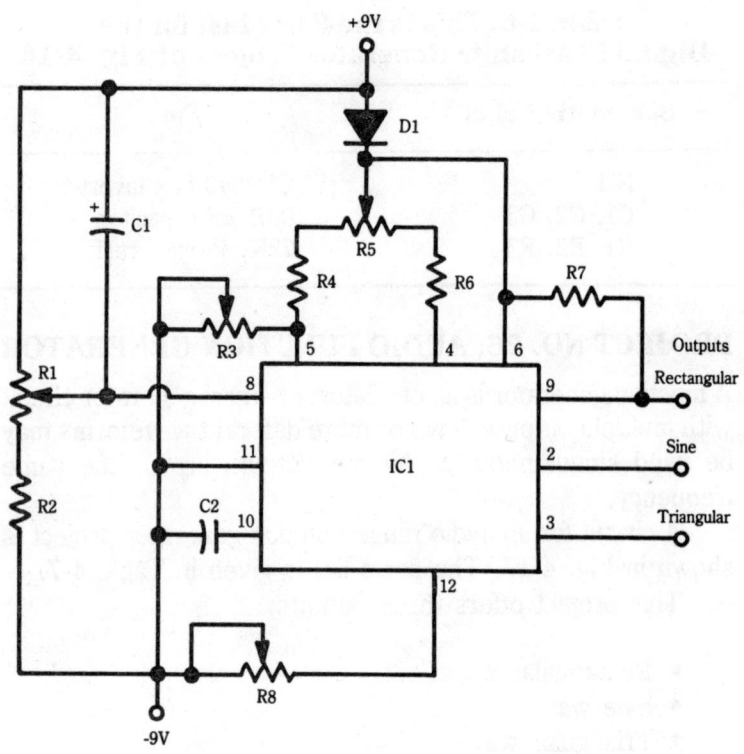

Fig. 4-17. This audio-function generator project puts out three basic waveforms. Project 26

Table 4-7. Parts List for the Audio Function Generator Project of Fig. 4-17.

Schematic Label	Part
IC1	8038 function generator IC
C1	1 µF, 35 volt electrolytic capacitor
C2	0.005 µF capacitor
R1	10K, ¼-watt resistor
R2	22K, ¼-watt resistor
R3	15 Megohm potentiometer
R4, R6	4.7K, ¼-watt resistor
R5	1K potentiometer
R7	18K, ¼-watt resistor
R8	100K trimpot

A wide output frequency range can be achieved with a single capacitance value.

The power supply requirements for the 8038 are quite flexible. Either a single-polarity or a dual-polarity power supply may be used. For a dual-polarity supply, the voltage should be in the range of ±5volts to ±15 volts. If a single-ended power supply is used, the voltage should be between +10 volts and +30 volts. For this project a dual-polarity power supply with a voltage of ±9 volts is used. You could power this circuit from a pair of standard 9 volt transistor batteries. If you choose to use batteries, the alkaline type is recommended.

The advantage of using a dual-polarity supply voltage is that the output waveforms will all be centered around ground potential (0 volts).

This project is designed to generate signals in the audible frequency range. The output frequency can be varied over a 1000:1 range, covering the complete spectrum of audible frequencies (20 Hz to 20 kHz). This wide range is achieved by applying different dc voltages to the FM sweep input (pin 8) via potentiometer R4. Meanwhile, the voltage across the timing resistors R1, R2, and R3 is held at a relatively low level by the 1N457 diode (D1). The voltage supplied to the timing resistors and the duty cycle control potentiometer (R3) is several millivolts below the maximum voltage available (V_{cc}) to the frequency control potentiometer (R4).

R8 is a trimpot. It is adjusted to minimize any variations in the duty cycle with changes in frequency. Trimpot R8 is adjusted for minimum sine wave distortion at a 50 percent duty cycle. An oscilloscope would certainly be helpful in making this adjustment, but if one is not available, feed the sine wave output signal to a small speaker and adjust R8 for the purest tone.

PROJECT NO. 27: SOUND-POCKET GENERATOR

This unusual project is a variation on both the basic astable multivibrator and the basic monostable multivibrator together. The circuit is shown in Fig. 4-18. The parts list is given in Table 4-8.

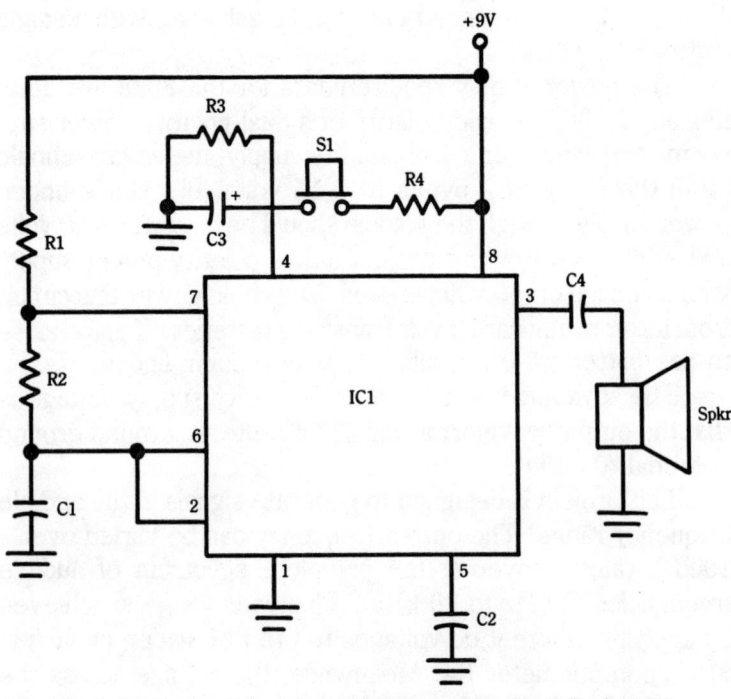

Fig. 4-18. This circuit is a "sound-pocket generator." Project 27

Table 4-8. Parts List for the Sound Pocket Generator Project of Fig. 4-18.

Schematic Label	Part
IC1	555 timer
C1, C4	0.1 µF capacitor
C2	0.01 µF capacitor
C3	50 µF, 25 Volt electrolytic capacitor
R1	2.2K, ¼-watt resistor
R2	22K, ¼-watt resistor
R3	120K, ¼-watt resistor
R4	3.9K, ¼-watt resistor
S1	Normally Open SPST push switch
SPKR	small speaker

The output is controlled by switch S1, which should be a momentary action Normally Open push-button type.

Initially, when power is applied, there is no output signal at all. When switch S1 is briefly closed, the circuit starts putting out a rectangular wave whose frequency and duty cycle are determined by the following formula:

$$F = 1[(1.1C1(R1 + R2)]$$

The speaker will continue sounding a constant tone as long as the switch is held closed. So far we have nothing special.

When the switch is released (opened), something interesting happens. The speaker will continue to produce the tone for a period defined by the time constant as follows:

$$T = 1.1R3C3$$

When C3 discharges, the output tone will cease.

In other words, each time S1 is momentarily closed, a "pocket" of sound at least as long as the R3C3 time constant will be produced. For want of a better name, I call this project a "sound pocket generator."

None of the component values are critical. Experiment with all of the parts values. Incidentally, you will notice that changing the value of R4 has no noticeable effect on circuit operation.

PROJECT NO. 28: TONE-BURST GENERATOR

This project could be considered an automated variation on the sound pocket generator circuit of Project No. 26. Again, the circuit is built around the two sections of the 556 dual timer. In this case, both timer sections are operated in the astable mode.

The circuit is shown in Fig. 4-19. A typical parts list is given in Table 4-9. Of course, you may experiment with other component values. This circuit's output will consist of bursts of tone, separated by periods of silence.

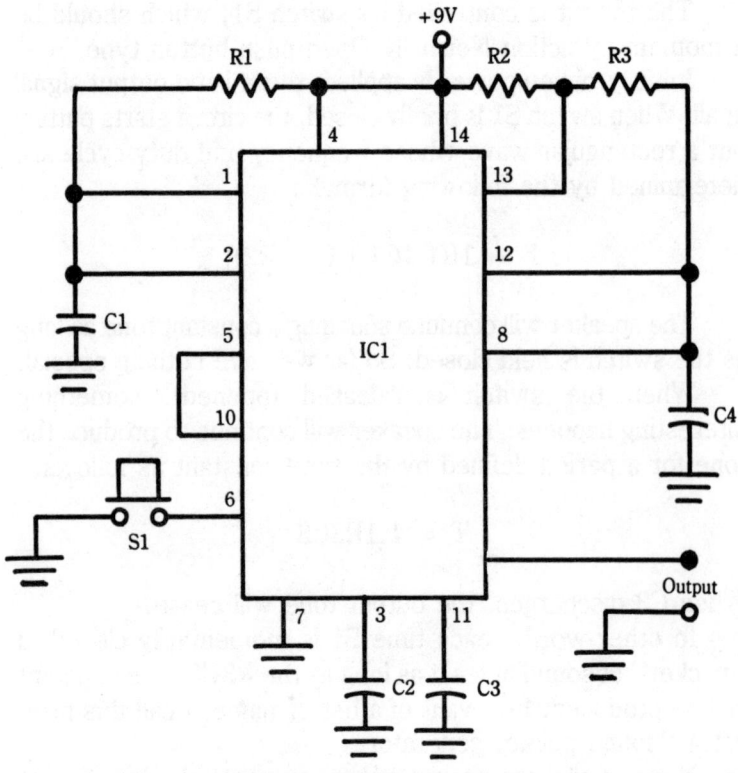

Fig. 4-19. This is the circuit for the tone-burst generator project. Project 28

Table 4-9. Parts List for the Tone Burst Generator Project of Fig. 4-19.

Schematic Label	Part
IC1	556 dual timer
C1, C4	0.1 µF capacitor
C2, C3	0.01 µF capacitor
R1	330K, ¼-watt resistor
R2	12K, ¼-watt resistor
R3	4.7K, ¼-watt resistor
S1	Normally Open SPST push switch

Basically, the first astable multivibrator stage turns the output of the second astable multivibrator stage on and off. The basic output waveform of this project is illustrated in Fig. 4-20.

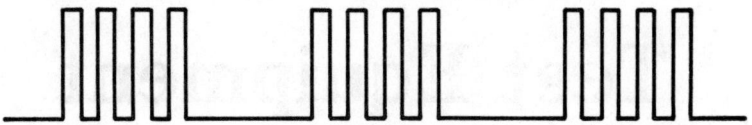

Fig. 4-20. This is the output signal from the circuit of Fig. 4-19.

An interesting effect can be achieved if both stages are set up for audible frequencies. The two signals will combine into a complex pattern, producing some very unusual tonal effects.

5
Test Equipment and Measuring Devices

Any electronics experimenter or technician has a frequent need for test equipment. Various electrical parameters such as voltage, current, or resistance need to be measured in many different circumstances.

Obviously, no one IC project is going to replace your digital multimeter or oscilloscope, but these simple projects are bound to come in very handy from time to time.

PROJECT NO. 29: AUDIBLE CONTINUITY TESTER

One of the most frequently made tests is the check for continuity. Is that solder joint making a good electrical connection? Is there a break in that connecting cable? Is that plug making proper contact with its jack? A continuity tester is used to answer all such questions.

You can use the ohmmeter section of your multimeter to make continuity tests, but that is a bit of overkill for such a simple test. It is often inconvenient to use a multimeter. You might need it in another part of the circuit. Besides, you have to look at the meter to determine the results, while at the same time watching to make sure the probes are where you want them. It's all too easy to let them slip out of position.

Fig. 5-1 shows a simple dedicated audible continuity tester circuit. When there is continuity between the two probes, a tone will sound. If there is a break in continuity, the speaker will remain silent.

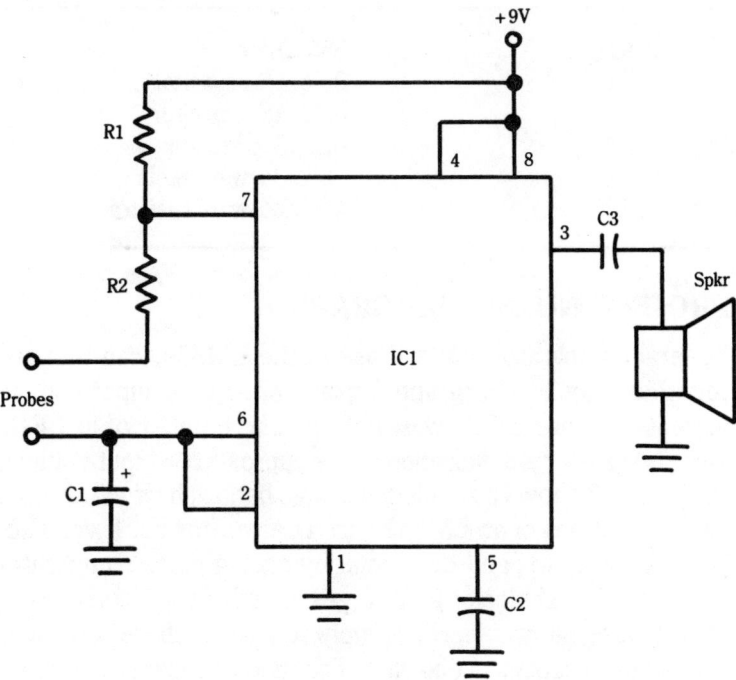

Fig. 5-1. This circuit gives an audible indication of continuity. Project 29

This circuit will also give you a rough indication of resistance. The lower the resistance between the probes, the higher the tone emitted by the speaker will be.

There is nothing complicated here. The project is simply a rectangular wave generator built around a 555 timer IC. The probes are placed in the circuit so that the resistance between them is part of the frequency determining resistance. If the resistance between the probes is very high, or infinite, no tone will sound. If the resistance is very high, but less than infinity, the circuit will be generating a signal, but it will be too low a frequency to be heard.

The parts list for this simple project is given in Table 5-1.

Table 5-1. Parts List for the Audible Continuity Tester Project of Fig. 5-1.

Schematic Label	Part
IC1	555 timer
C1	0.05 µF capacitor
C2	0.01 µF capacitor
C3	0.1 µF capacitor
R1	68K, ½-watt resistor
R2	47K, ½-watt resistor

PROJECT NO. 30: BARGRAPH

Several comparators, like those in the LM339, can be used together to build a bargraph display unit. As the input voltage increases, more LEDs will light up. The number of lit LEDs gives an easily read indication of the approximate input voltage.

Fig. 5-2 shows a simple four stage bargraph circuit. It uses all four sections of an LM339 quad comparator IC. If you like, you can easily expand the circuit by adding more comparator sections. The choice of four stages is entirely arbitrary. Four stages are shown here, simply because there are four comparator sections in one chip. The more comparator sections you use, the wider the range of the circuit.

The value of resistor R1 determines the sensitivity of the circuit. You could use a potentiometer here as a calibration control.

Resistors R2 through R5 should have identical values for equal steps per LED. You could use unequal resistor values to weight the scale for special applications.

Resistors R6 through R9 simply protect the LEDs from excessive current flow. The values of these resistors determine the brightness of their associated LEDs.

The only other thing to bear in mind for this project is that the input signal is not referenced directly to ground. The input voltage is connected across the two points marked in the diagram.

A typical parts list for this project is given in Table 5-2.

Table 5-2. Parts List for the Bargraph Project of Fig. 5-2.

Schematic Label	Part
IC1	LM339 quad comparator
D1–D4	LED
R1	68K, ¼-watt resistor
R2–R5	2.2K, ¼-watt resistor
R6–R9	330 ohm, ¼-watt resistor

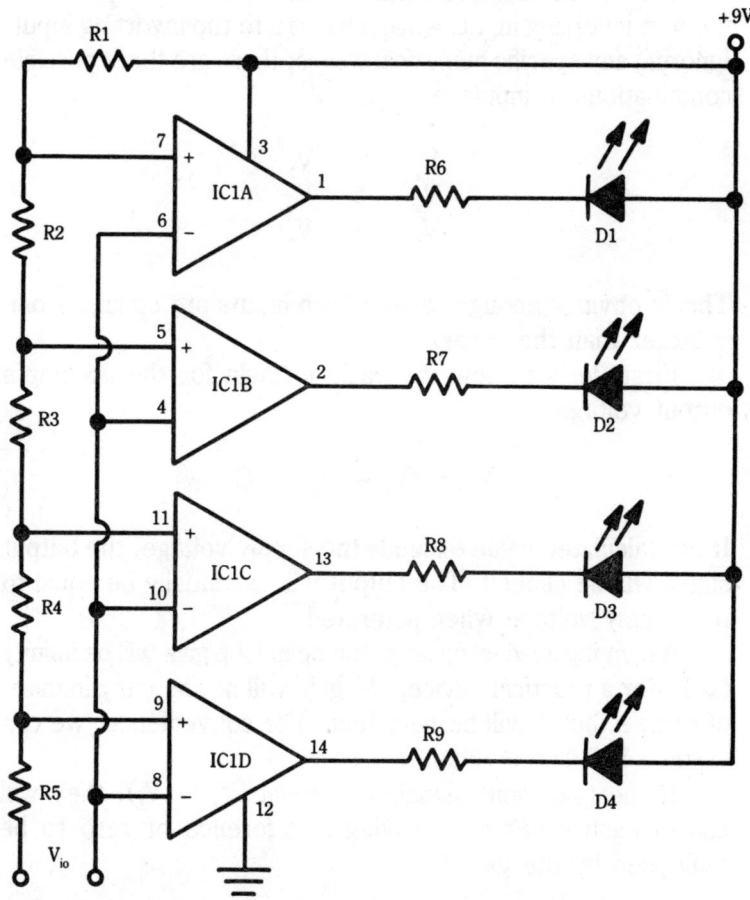

Fig. 5-2. Multiple comparator stages are used to create a bargraph. Project 30

PROJECT NO. 31: VOLTAGE COMPARATOR

The comparator is a very aptly named circuit. It actually compares two input signals and tells you which one is larger, assuming both signals have the same polarity.

Comparators are generally used in single polarity applications. We will make that assumption throughout our discussion.

A comparator is basically an op amp with no feedback loop. The op amp operates with its full open loop gain, which is very high. Theoretically, the gain is infinite.

We will call the two input signals V_a and V_b. V_a is fed to the non-inverting input, while Vb is fed to the inverting input. Ignoring any specific numerical values, there are three possible combinations of inputs:

$$V_a = V_b$$
$$V_a > V_b$$
$$V_a < V_b$$

This is obvious enough. Either both inputs are equal, or one is larger than the other.

First, let's review the basic formula for the op amp's output voltage:

$$V_o = (V_a - V_b) \times G$$

If the calculated value exceeds the supply voltage, the output signal will be clipped. The output will essentially be equal to the supply voltage when saturated.

Assuming an ideal op amp, the open loop gain will be infinity (∞). For a practical device, the gain will not be truly infinite, of course, but it will be very high. For convenience, we can assume the gain is infinite.

If the two input signals are equal ($V_a = V_b$), they will cancel each other out, leaving a difference of zero to be multiplied by the gain:

$$V_o = (V_a - V_b) \times G$$
$$= 0 \times G$$
$$= 0$$

Multiplying anything, even a very, very large value, by zero always leaves zero. Whenever the output of a comparator is zero, then you know the two input voltages must be equal.

Now, what happens if V_a is larger than V_b, even if only by a very small amount? Remember, we are assuming that both input voltages are positive. If V_a is larger than V_b, then the difference value will be positive. Even a tiny positive value multiplied by a large gain will be positive and very large. Multiplying even a small positive value by the large open-loop gain will saturate the output. If V_a is larger than V_b, the output will be approximately the positive supply voltage.

On the other hand, if V_a is smaller than V_b, then the difference value will be negative, and the output will be saturated in the negative direction.

To summarize, there are only three possible output conditions, each uniquely defining the relationship of the two input voltages:

Output	Input Condition
0	$V_a = V_b$
V+	$V_a > V_b$
V-	$V_a < V_b$

The differences between these output states are very clearcut and obvious. Further, they are easy to detect by other circuitry.

In most practical comparator applications, one of the two input voltages will be fixed. This is called the reference voltage. A variable voltage is fed to the second input, and is compared to the reference voltage.

A practical comparator project is illustrated in Fig. 5-3. The parts list is given in Table 5-3.

R1 through R3 is a simple resistive voltage divider network. By adjusting potentiometer R2, you can manually control the input voltage. To use the circuit with an external

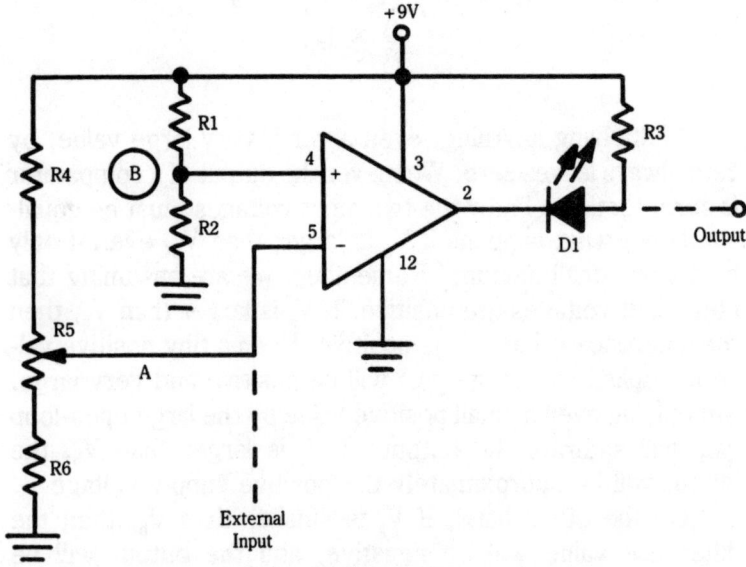

Fig. 5-3. This is a practical comparator project. Project 31

**Table 5-3. Parts List for the
Voltage Comparator Project of Fig. 5-3.**

Schematic Label	Part
IC1	LM339 quad comparator
D1	LED
R1, R2	10K, ¼-watt resistor
R3	470 ohm, ¼-watt resistor
R4, R6	4.7K, ¼-watt resistor*
R5	10K potentiometer*

*eliminate if external input is used

signal, simply eliminate R1 through R3, and feed the input signal into R4.

In this project, the reference voltage is O. The non-inverting input is connected to the circuit ground through R5. You can feed any desired voltage in as the reference value. With a reference voltage of O, the comparator circuit will effectively indicate the polarity of the input voltage.

The output state is indicated by two LEDs. Only one LED should ever be lit at any given time. If both LEDs are lit, there is something wrong with the circuit. Recheck your wiring.

The three output conditions are indicated as follows:

Both LEDs dark	$V_a = V_b$
LED1 lit - LED2 dark	$V_a > V_b$
LED1 dark - LED2 lit	$V_a < V_b$

While experimenting with this project, you might find a point where one of the LEDs is dimly lit. This does not indicate a problem. The output is not completely saturated. This can happen because the open loop gain of a practical op amp is finite. If the two input voltages are unequal by only a very small amount, the output may not be fully saturated to the clipping point. A graph of the comparator's output is shown in Fig. 5-4. This "range of uncertainty" is, fortunately, very small, and can almost always be ignored without much concern about unreliable operation.

PROJECT NO. 32: VOLTAGE RANGE DETECTOR

Sometimes it is necessary to know if a voltage is within a specific range. You could wire in a voltmeter, of course, but in many applications that would be overkill. If you are only concerned with whether or not the voltage is within the specified range, the exact voltage isn't important. Reading a voltmeter is not really difficult, but it does take some effort. Often it will be handy to have a simple go/no go LED indicate whether or not the voltage is in the acceptable range. That is the purpose of this project. The circuit is shown in Fig. 5-5.

This project is designed around two sections of an LM339 quad comparator. Op amps can be used as comparators, but the LM339 is specifically designed for such applications.

The technical name for this project is a "limit comparator." The LED at the output lights up only when the input voltage is within the specified range, determined by the values of R1, R2, and R3. If the input voltage goes even slightly below or

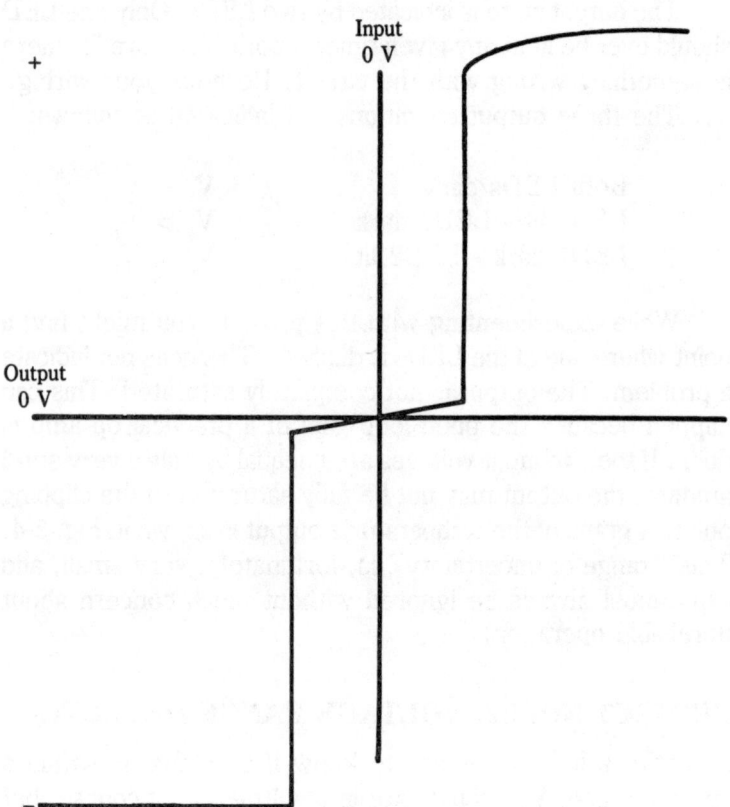

Fig. 5-4. The comparator circuit has a small "range of uncertainty."

above the limits of the specified range, the LED will be extinguished.

The "on" range is set by resistors R1, R2, and R3, which are arranged as a simple voltage divider network. R1 sets the upper cut-off limit, and R3 sets the lower cut-off limit. Since R2 is between the upper limit comparator and the lower limit comparator, its value will determine how wide the "on" range is. A small value for R2 will give the circuit a fairly narrow range of input voltages that will turn on the LED. Conversely, if a large value resistor is used, the "on" range will be proportionately wider.

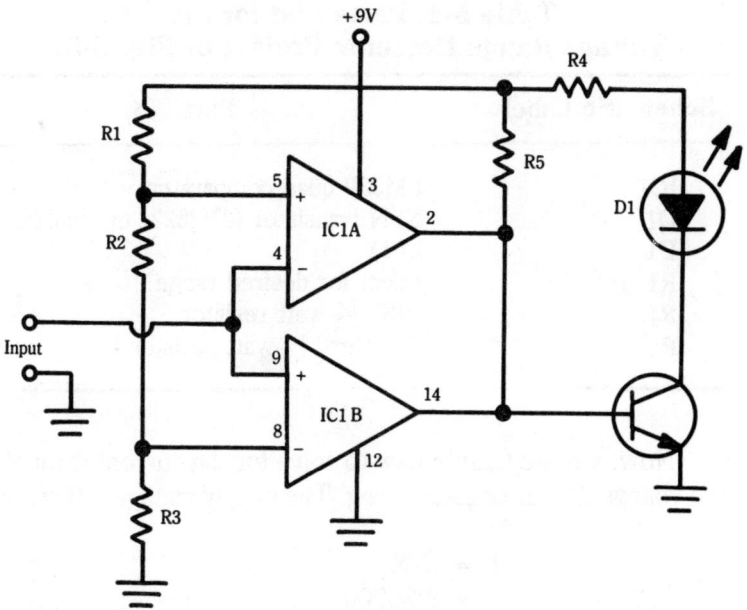

Fig. 5-5. This advanced comparator project indicates whether or not the input voltage is within a specified range. Project 32

The parts list for this project is given in Table 5-4. Notice that the values for R1, R2, and R3 are not listed. They should be selected for your specific application.

The limit setting resistors are selected using Ohm's Law. As an example, let's assume the following resistors are used:

$$R1 = 10K$$
$$R2 = 4.7K$$
$$R3 = 22K$$

The total resistance in the string is therefore equal to:

$$Rt = 10000 + 4700 + 22000$$
$$= 36,700 \text{ ohms}$$

Table 5-4. Parts List for the Voltage Range Detector Project of Fig. 5-5.

Schematic Label	Part
IC1	LM339 quad comparator
Q1	NPN transistor (2N2222, or similar)
D1	LED
R1, R2, R3	select for desired range
R4	10K, ¼-watt resistor
R5	330 ohm, ¼-watt resistor

Now, we use Ohm's Law to solve for the current through the voltage divider resistor string. The supply voltage is 9 volts:

$$\begin{aligned} I &= E/R \\ &= 9/36700 \\ &= 0.0002452 \text{ amp} \\ &= 0.2452 \text{ mA} \end{aligned}$$

The current flow is equal for each of the resistors in the string. This allows us to determine the voltage drop across each of the resistors:

$$\begin{aligned} E &= IR \\ E1 &= 0.0002452 \times 10000 \\ &= 2.452 \text{ volts} \\ E2 &= 0.0002452 \times 4700 \\ &= 1.15244 \text{ volt} \\ E3 &= 0.0002452 \times 22000 \\ &= 5.3944 \text{ volts} \end{aligned}$$

The total voltage drops should equal the original supply voltage (9 volts):

$$\begin{aligned} E &= E1 + E2 + E3 \\ &= 2.452 + 1.15244 + 5.3944 \\ &= 8.99884 \text{ volts} \end{aligned}$$

The small error is due to rounding off values in the previous equations.

Resistor R3 sets the lower cut-off limit at approximately 5.4 volts. R1 sets the upper cut-off limit at E − E1 volts. In this case:

$$9 - 2.5 \cong 6.5 \text{ volts}$$

The "on" range is about 1.1 volt wide. Experiment with other resistor values.

PROJECT NO. 33: LOGIC PROBE

A simple dc voltmeter can be used to monitor digital signals, but it really isn't designed for the job. A more elegant way to test digital signals is to use a specialized piece of test equipment called a "logic probe."

Logic probes are simple but powerful tools for analyzing what goes on in a digital electronics circuit. Many commercially available logic probes with all sorts of extra features can be purchased, but this project will probably come in very handy for the digital electronics experimenter. What this device lacks in special features, it more than makes up for with low cost. The project shouldn't cost you more than two or three dollars, even with all new components.

A logic probe is simply a device that indicates the logic state, 0 or 1, at any point within a digital circuit. For illustration purposes, a super-simple two component logic probe is shown in Fig. 5-6. The ground lead can be fitted with an alligator clip so it can be connected to the same ground as the circuit being tested.

The probe is just a short length of stiff solid wire, or a common test lead probe. If this probe is touched to a point in the circuit with a logic 1 (High) signal, the LED will light up. At all other times, the LED will remain dark.

The resistor is included for current limiting. If the LED is permitted to draw too much current it can be damaged. This current limiting resistor will generally have a value of less than 1K (1000 ohms), but more than 100 ohms. A typical value is

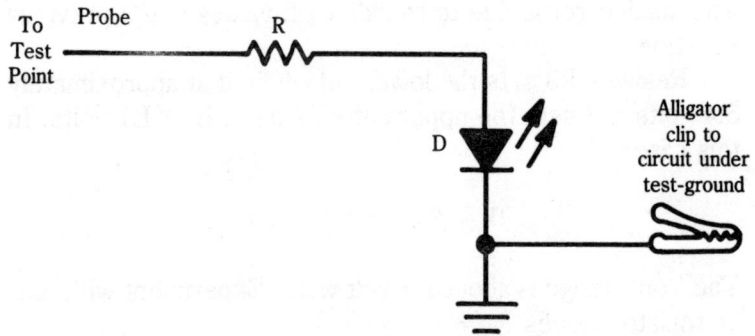

Fig. 5-6. A simple logic probe can be made from nothing more than an LED and a resistor.

330 ohms. The smaller the resistance, the brighter the LED will glow.

While functional, this super-simple approach to a logic probe definitely leaves a lot to be desired. For one thing, the circuit does not give a definite indication of a logic 0 (Low) state. If the LED does not light up, you *may* have a logic 0 signal, or the probe may not be making good contact with the pin being tested. There might even be a broken lead, or the LED itself could be damaged. With this circuit, there is simply no way to tell.

Another problem with this circuit is that it is entirely passive and parasitic. It can excessively load down some circuits, causing incorrect operation. Obviously, a piece of test equipment that causes the circuit to malfunction isn't going to give you any meaningful readings.

Loading problems are most likely to occur in complex systems where the gate being tested is already driving close to its maximum fan-out potential.

Both of these problems are sidestepped in the actual logic probe project, which is shown in Fig. 5-7. The parts list for this project is given in Table 5-5.

The chief difference here is that a pair of inverter stages is used, making this an active, rather than a passive device.

When a logic 0 signal is applied to the probe tip, the first inverter will change it to a logic 1, lighting LED A. The second inverter changes the signal back to a logic 0, so LED B

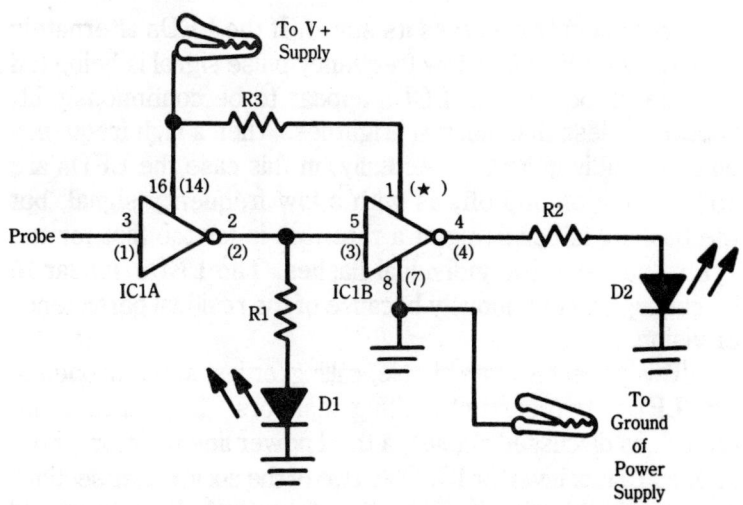

Fig. 5-7. A practical logic probe indicates High, Low, and pulse conditions. Project 33

Table 5-5. Parts List for the Logic Probe Project of Fig. 5-7.

Schematic Label	Part
IC1	CD4009A hex inverter
D1, D2	LED
R1, R2	330 ohm, ¼-watt resistor
R3	1K, ¼-watt resistor

remains dark. Conversely, when a logic 1 signal is fed into the probe tip, the output of the first inverter stage is a logic 0, so LED A stays dark, while the second inverter changes the signal back to a logic 1, causing LED B to light up.

As you can see, this circuit can give a definite indication of either a logic 1 or a logic 0 condition. If neither LED lights up, the probe is not making proper contact, or the test point is dead (no signal). This circuit is much less ambiguous than the earlier version.

But that's not all this logic probe project can tell you. This device can also indicate the presence of a pulse signal (a signal

109

that continuously reverses its state). If the LEDs alternately blink on and off, then a low frequency pulse signal is being fed into the probe. If both LEDs appear to be continuously lit, possibly at less than normal brightness, then a high frequency pulse signal is indicated. Actually, in this case the LEDs are still blinking on and off, as with a low frequency signal, but the blinking is occurring at a rate that is far too fast for the eye to perceive the individual flashes. The LEDs appear to be staying on continuously because of the residual persistence of vision.

This project is quite simple, calling for just six components: two LEDs, two current limiting resistors (their value is not critical, as discussed above), a third power line resistor (about 1K), and a hex inverter IC. Only two of the six inverter sections are used in this circuit. The other four unused inputs should be grounded to insure circuit stability. This is particularly important if a CMOS chip is being used, as called for in the parts list.

Like the simple logic probe described earlier, this circuit "steals" its power from the circuit being tested. Alligator clips on the power supply leads (V+ and ground) can be attached to the power supply output of the circuit being tested.

Almost any inverters may be used in this project. A CMOS CD4009A hex inverter IC is the best choice, since it can be driven by either TTL or CMOS gates. By tapping into the tested circuit's power supply, the voltage levels will be automatically matched. In some cases, it may be desirable to add a pull-up resistor at the probe's input, as shown in Fig. 5-8. This will only be necessary if you have some reliability problems in measuring signals in TTL circuits. If used, the value of this pull-up resistor is not critical. It should be around 1K, but anything reasonably close will do.

If you intend to use the logic probe with TTL circuits only, you can substitute a 7404 hex inverter chip, or the equivalent in the appropriate sub-family, like the low power Schottky 74LS04. Of course, the pin numbers will be changed, so check the data sheet or a data book.

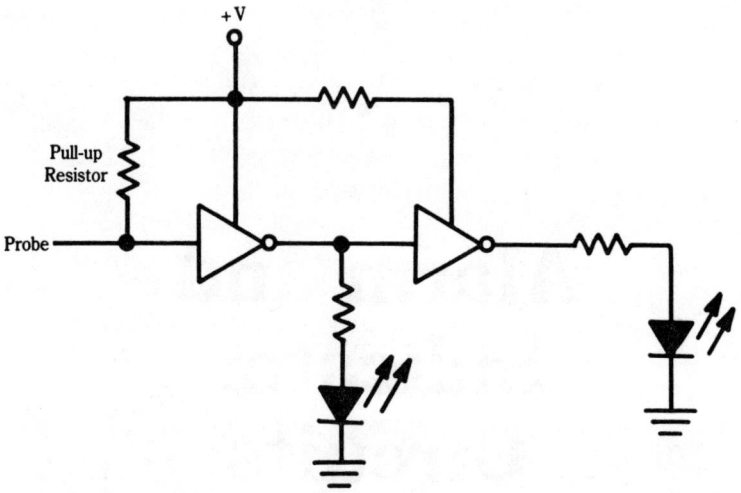

Fig. 5-8. In some applications a pull-up resistor may be added to the project's input probe.

The pin numbers in parentheses in the schematic are for the 7404. The pin marked * (pin No. 1) has no equivalent on the 7404, and should simply be ignored.

6

Alarm and Indicator Circuits

The projects in this chapter are intended to detect a specific condition and produce a warning, so the condition may be corrected.

PROJECT NO. 34: SIMPLE BURGLAR ALARM

Unfortunately, in today's world almost everyone needs a burglar alarm of some kind. There have been many elaborate systems designed over the years. Some are very sophisticated and virtually unbeatable, but no alarm system is ever 100 percent unbeatable. However, such systems are quite expensive.

When a lower level of security is acceptable, a simple home brew alarm system will do a good job at a much lower cost.

In simple alarm systems, special switching devices are used to monitor possible entry points, such as doors and windows. Foil tape on a window will break if the window is broken, opening the monitor circuit and triggering the alarm.

Many doors and windows can be monitored by magnetic reed switches. These switches are in two sections: one section, mounted on the door as illustrated in Fig. 6-1, contains a small permanent magnet; the other section, mounted on the doorjamb, contains the actual switching element and con-

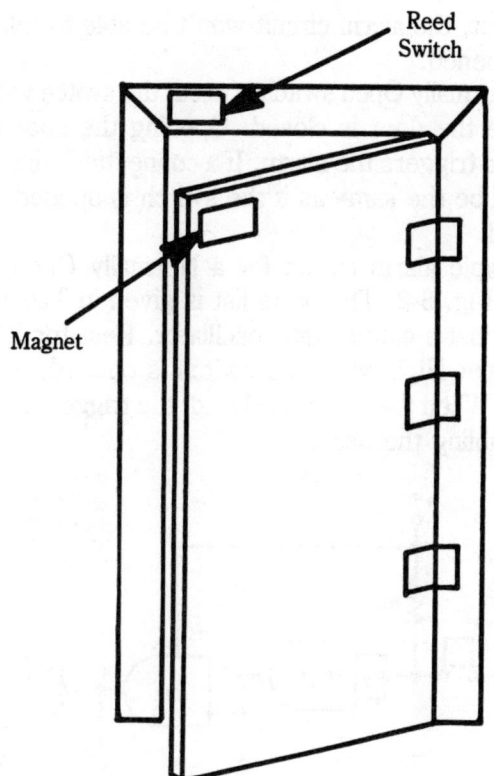

Fig. 6-1. Magnetic reed switches are often used in intrusion detection applications.

nections to wire it into the circuit. The switch is activated when the magnet is brought close to the switching element. The two sections are mounted so the switch is activated when the door is closed.

Magnetic reed switches are available in both Normally Open and Normally Closed versions.

A Normally Closed switch will look like an open circuit when the door is closed. If the door is opened, the switch contacts are closed. This might seem like the obvious choice, because the alarm triggering circuit is very simple. Closing the switch turns the alarm device on. Unfortunately, a Normally Closed set-up is very easy to defeat. If one of the connecting wires is cut, the alarm circuit won't be able to tell when the door is opened.

If a Normally Open switch is used, the switch will be closed only while the door is closed. Opening the door opens the switch and triggers the alarm. If a connecting wire is cut, the effect will be the same as if the switch is opened: the alarm will sound.

A simple alarm circuit for a Normally Open system is shown in Fig. 6-2. The parts list is given in Table 6-1. This project is just a gated digital oscillator. Resistor R1 pulls the trigger input High when the switch is opened, enabling the oscillator. When the switch is closed, the trigger input is forced Low, disabling the oscillator.

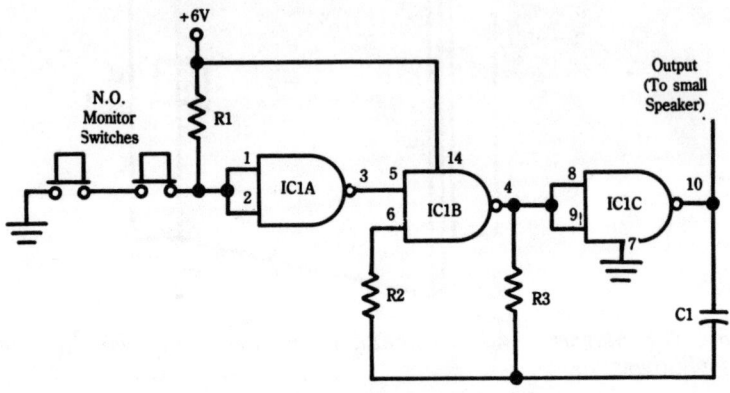

Fig. 6-2. This is a simple burglar alarm circuit. Project 34

Table 6-1. Parts List for the Simple Burglar Alarm Project of Fig. 6-2.

Schematic Label	Part
IC1	CD4011 quad NAND gate
C1	0.01 μF capacitor
R1, R2	1 Megohm, ½-watt resistor
R3	120K, ½-watt resistor

Multiple Normally Open switches can be monitored with this circuit. Just wire them in series.

PROJECT NO. 35: FLOODING ALARM

This is another useful project for the homeowner, especially if the homeowner is plagued with a leaky basement. It can be used to protect anything that must be kept dry.

The schematic diagram for the Flooding Alarm project is shown in Fig. 6-3. The parts list is given in Table 6-2.

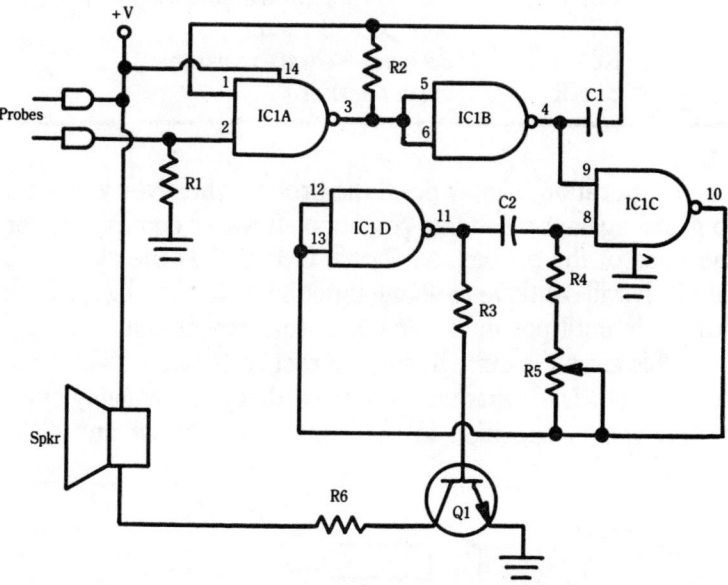

Fig. 6-3. This alarm circuit sounds a warning when a flooding condition is detected. Project 35

This project takes advantage of the fact that water is conductive. The probes are placed so that unwanted water or other liquid touches both probes, creating a short between them. When this happens, the alarm will sound.

Potentiometer R5 determines the frequency, or pitch of the alarm tone. A fixed resistance may be substituted, if you prefer.

Table 6-2. Parts List for the Flooding Alarm Project of Fig. 6-3.

Schematic Label	Part
IC1	CD4011 quad NAND gate
Q1	NPN transistor (2N3904, or similar)
C1	0.1 µF capacitor
C2	0.01 µF capacitor
R1	1 Megohm, ¼-watt resistor
R2	3.3 Megohm, ¼-watt resistor
R3, R4	10K, ¼-watt resistor
R5	50K potentiometer
R6	33 ohm, ¼-watt resistor
SPKR	small speaker

In operation, simply place the probes wherever you want to guard against a flooding condition. If water comes up over the ends of the probes, as shown in Fig. 6-4, the alarm will go off. It will continue to sound until the water level goes back down, or until power is removed from the circuit.

This circuit is quite flexible. Exact component values are not needed. Use whatever you have that is reasonably close to the values specified in the parts list. Almost any NPN

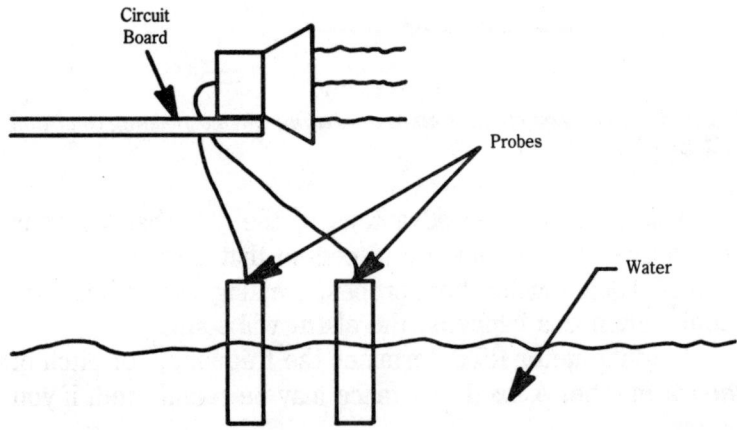

Fig. 6-4. Excess liquid provides a short circuit between the two probes.

transistor may be used for Q1. Even the power supply requirements for this project are quite flexible. The supply voltage may be anything from +6 to +12 volts.

PROJECT NO. 36: LIGHT-ON ALARM

Sometimes it is helpful to know when a light is on. For example, you could mount a sensor in your garage to warn you when you forget to turn off your headlights.

A simple CdS photoresistor can be added to a basic 555 astable multivibrator circuit to create a light on alarm. The circuit is shown in Fig. 6-5. The parts list is given in Table 6-3.

When the sensor is dark, the speaker will be silent; but once the light level on the sensor exceeds a critical level, a tone will be sounded.

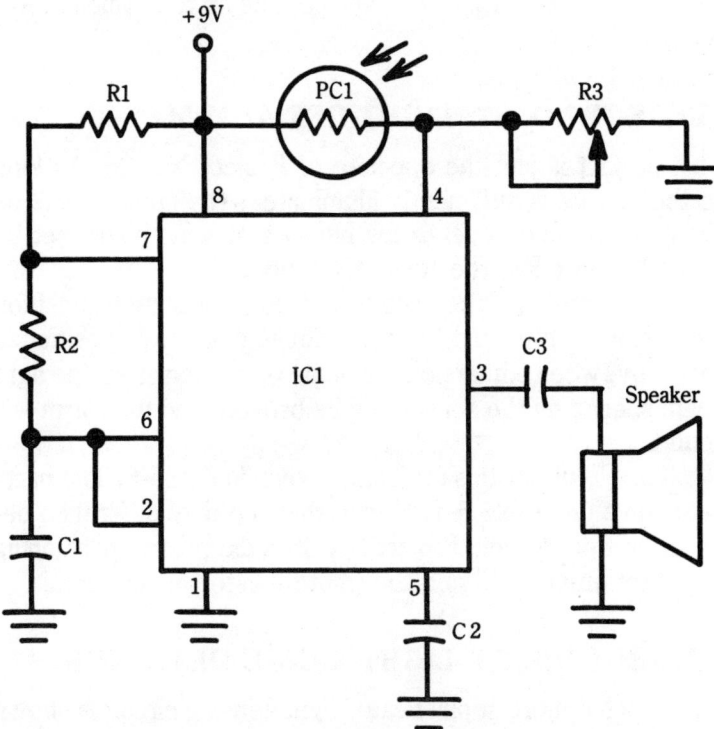

Fig. 6-5. This circuit warns you when a light is on. Project 36

**Table 6-3. Parts List
for the Light On Alarm Project of Fig. 6-5.**

Schematic Label	Part
IC1	555 timer
PC1	Cds photoresistor
C1, C2	0.01 µF capacitor
C3	0.1 µF capacitor
R1	100K, ¼-watt resistor
R2	2.2K, ¼-watt resistor
R3	10K potentiometer

Potentiometer R3 is used to adjust the sensitivity of the circuit. A fixed resistor may be substituted if you prefer. The frequency of the output tone will depend on the values of R1, R2, and C1.

PROJECT NO. 37: LIGHT-OFF ALARM

This project is just the opposite of Project No. 35. As long as the sensor is sufficiently illuminated, the speaker will be silent. If the light level drops below a specific point, set by potentiometer R3, the tone will sound.

This circuit may be used for intrusion detection. Position the sensor with a light source shining on it. If positioned correctly, when someone enters the protected area, the light beam shining on the sensor will be broken, and the alarm will sound.

The circuit for this project is shown in Fig. 6-6. The parts list is given in Table 6-4. Notice that the only difference between this project and Project No. 35 is the relative positioning of potentiometer R3 and the photoresistor in the circuit.

PROJECT NO. 38: LIGHT-RANGE DETECTOR

A somewhat more sophisticated light sensing circuit is shown in Fig. 6-7. The parts list for this project appears in Table 6-5.

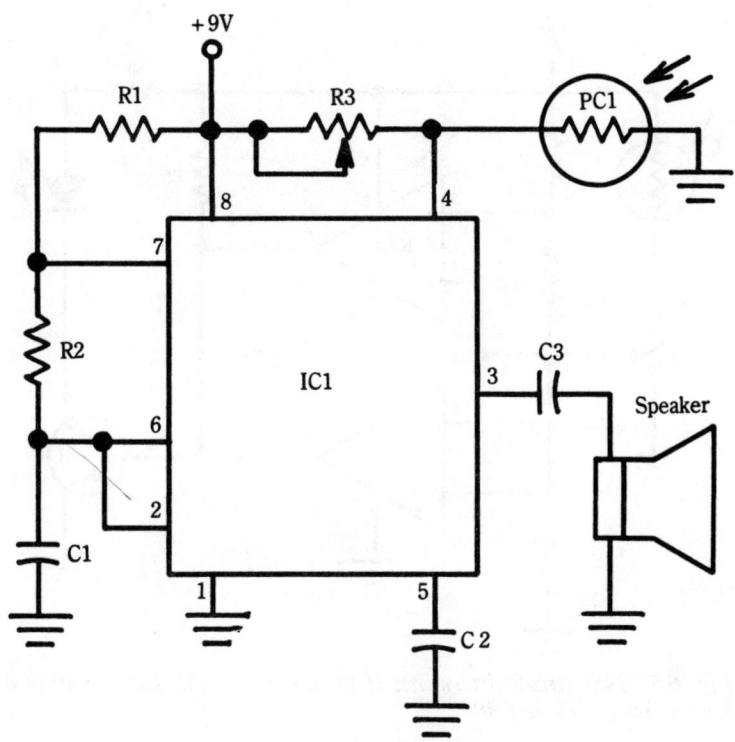

Fig. 6-6. This alarm circuit sounds a warning when it is in the dark. Project 37

Table 6-4. Parts List for the Light Off Alarm Project of Fig. 6-6.

Schematic Label	Part
IC1	555 timer
PC1	Cds photoresistor
C1, C2	0.01 µF capacitor
C3	0.1 µF capacitor
R1	100K, ¼-watt resistor
R2	2.2K, ¼-watt resistor
R3	10K potentiometer

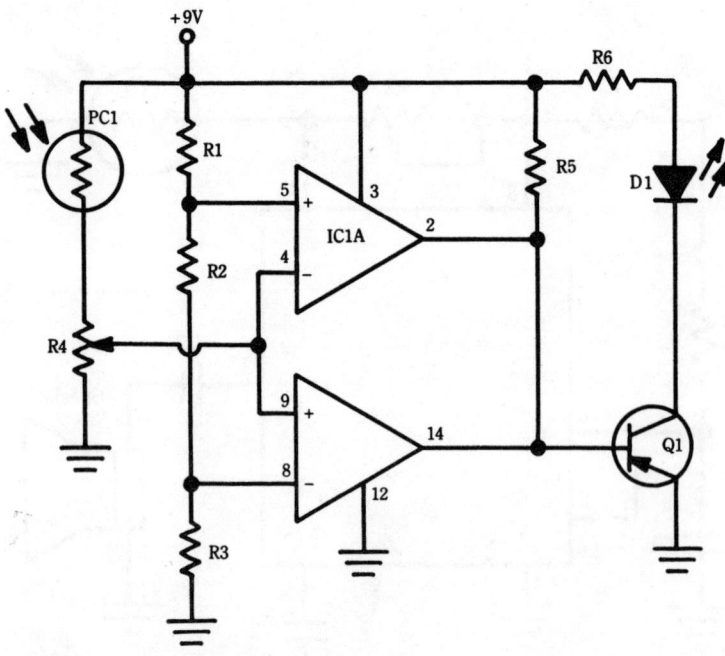

Fig. 6-7. This circuit determines if the ambient light level is within a specific range. Project 38

Table 6-5. Parts List for the Light Range Detector Project of Fig. 6-7.

Schematic Label	Part
IC1	LM339 quad comparator
Q1	PNP transistor (2N3906, or similar)
D1	LED
PC1	photoresistor
R1, R2, R3	select for desired range
R4	100K potentiometer
R5	10K, ¼-watt resistor
R6	330 ohm, ¼-watt resistor

This is a comparator circuit that gives an output indication (the LED is lit) if the light striking the sensor is within a specific range. If the light level is too high or too low, the LED will be dark.

The most obvious applications for this circuit are in photography, where light levels are often critical.

The input to this comparator is tapped off from a simple two part voltage divider made up of the photoresistor (PC1) and a potentiometer (R4). Each of these components has a variable resistance. R4 varies with its manual setting, so it functions as a range control. The resistance of PC1 depends on the intensity of the light striking it.

If the light striking the photocell is within a given range, determined by the values of R1, R2, and R3, the output goes High, and the LED lights up.

Experiment with various values of R1, R2, and R3. A fixed resistor may be substituted for R4, if your application doesn't require a manual range control.

7
Filters

Most ac signals contain multiple frequency components. Only a pure sine wave consists of a single frequency. In a number of applications, we will need to remove certain frequency components from the signal. For this purpose, a frequency sensitive circuit known as a filter is used.

There are four basic types of filters. The frequency response graphs for each type are illustrated in Figs. 7-1 through 7-4.

A low-pass filter removes the upper frequency components, while leaving the lower frequency components alone (Fig. 7-1).

A high-pass filter (Fig. 7-2) is just the opposite of a low-pass filter. Low frequency components are blocked, while high frequency components are permitted to pass on through to the output.

A band-pass filter (Fig. 7-3) blocks all frequency components, except those within a specific range, or band.

A band-reject filter (Fig. 7-4) is just the opposite of a band-pass filter. Only those frequency components within the specified band are blocked. All other frequency components are passed.

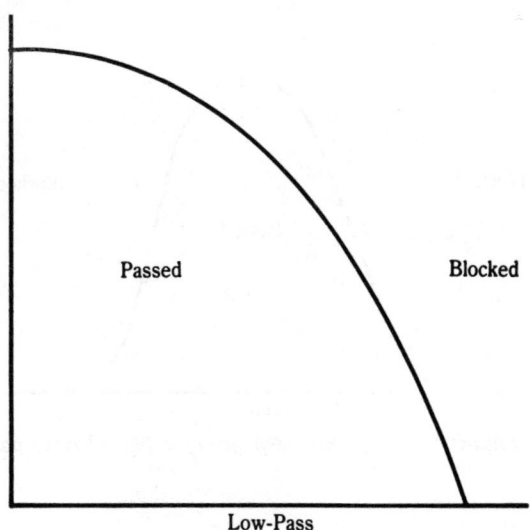

Fig. 7-1. This is the frequency response graph for a low-pass filter.

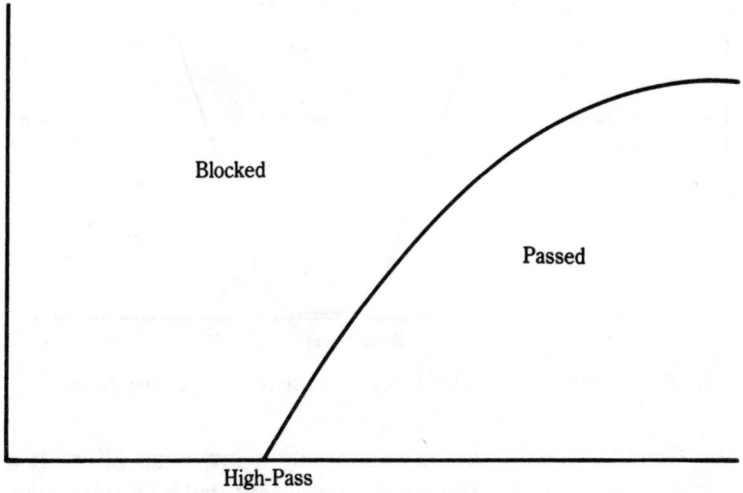

Fig. 7-2. This is a frequency response graph for a high-pass filter.

PROJECT NO. 39: INTEGRATOR

The circuit for an integrator project is shown in Fig. 7-5. This circuit performs the mathematical operation known as integration. Unless you've studied calculus, you're probably not familiar with integration. So what is this project good for?

123

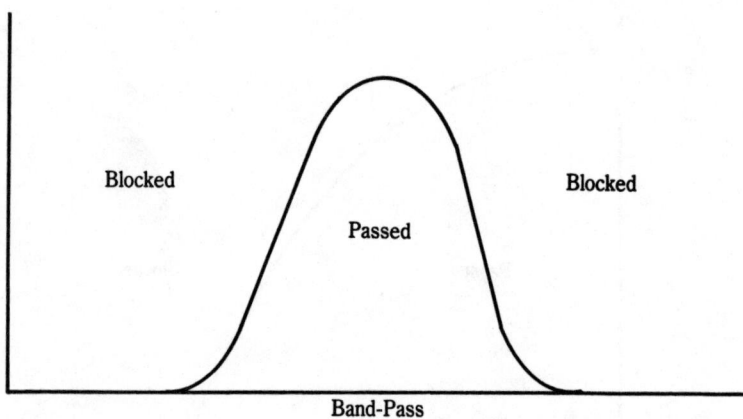

Fig. 7-3. This is the frequency response of a typical band-pass filter.

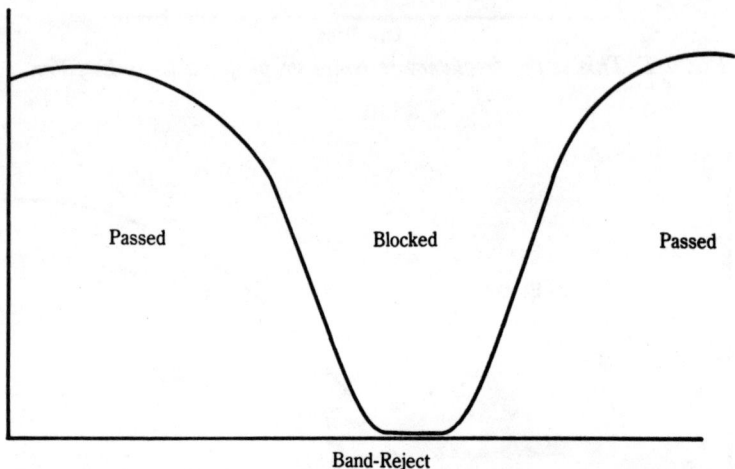

Fig. 7-4. A band-reject filter is the opposite of a band-pass filter.

Calculus aside, this is essentially a low-pass filter. If a square wave is fed in the input, the output will be a triangular wave. If a triangular wave is the input, the output will resemble a sine wave.

For serious audio applications, you will probably want to use a high grade low noise op amp in place of the 741 op amp listed in the parts list.

A recommended parts list for this integrator project is given in Table 7-1. Experiment with other component values.

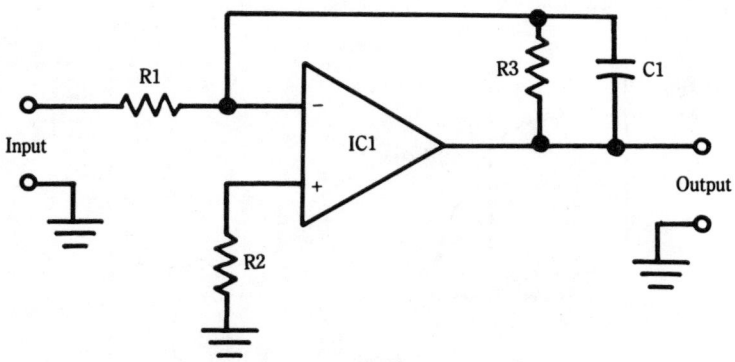

Fig. 7-5. An integrator is a form of low-pass filter. Project 39

Table 7-1. Parts List for the Integrator Project of Fig. 7-5.

Schematic Label	Part
IC1	op amp (741, or similar)
C1	0.1 μF capacitor
R1, R2	10K, ¼-watt resistor
R3	100K, ¼-watt resistor

PROJECT NO. 40: ACTIVE BAND-PASS FILTER

A filter is a frequency sensitive circuit. It allows some frequencies to pass through from input to output with little or no attenuation. In some cases the passed frequency components are actually amplified. Other frequencies are greatly attenuated, or (ideally) blocked completely from the output.

Filters are used to remove undesired harmonics, or other frequency components from a complex signal. There are four basic types of filters. Their names pretty much tell the story. All four types of filtering are illustrated in Fig. 7-6.

Low-Pass Filter

Low frequencies are passed, and high frequencies are blocked. The point of division between passed and blocked is

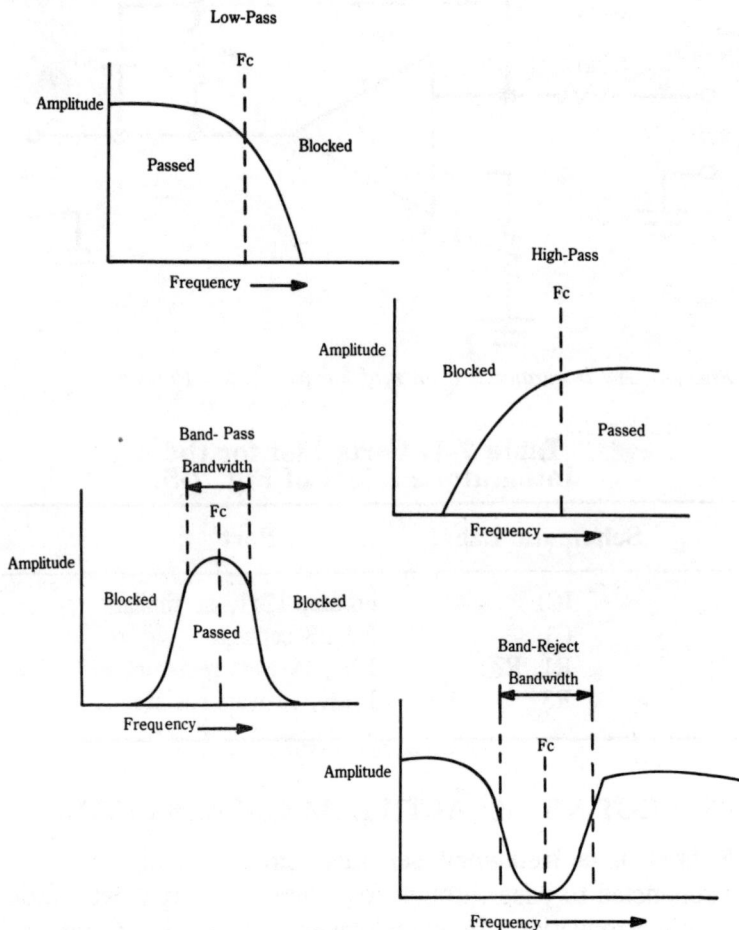

Fig. 7-6. The four basic types of filters are compared here.

called the cut-off frequency. Note that this is not a sharp angle, but a point on a smooth curve. The steeper the curve, the better the filter.

High-Pass Filter

This is the opposite of the low-pass filter. This time the low frequencies are blocked, and high frequencies are passed.

If the input is a complex waveform, the fundamental will be deleted, and you will be left with the upper frequencies at the output.

Band-Pass Filter

This type of filter passes only those frequencies within a specific range, or band. Any frequency component outside this specified band is blocked. A band-pass filter has two important specifications. The center frequency is the mid-point of the passed band of frequencies. The other important specification is the bandwidth, or the width of the range of passed frequencies.

Band-Reject Filter

This is the opposite of the band-pass filter. All frequency components are passed, except those within a specified band, which are blocked. The band-reject filter is normally used to remove a specific interference signal, such as 60 Hz hum from ac power lines. This type of filter is also known as the "notch filter."

This project is an active band-pass filter with a narrow bandwidth. The schematic for this project appears in Fig. 7-7. Figure 7-8 shows the output graph for this circuit.

The passed band centers around a specific frequency (Fc). This center frequency is defined by the values of the capacitor and the coil in the feedback loop, according to the formula:

$$F_c = 1/(2\pi\sqrt{LC})$$

where π is the mathematical constant *pi*, with a value of approximately 3.14.

The gain for the passed frequencies is defined by the ratio of the input (R1) and feedback (R2) resistances, as in a standard inverting amplifier:

$$V_o = V_i \times R2/R1$$

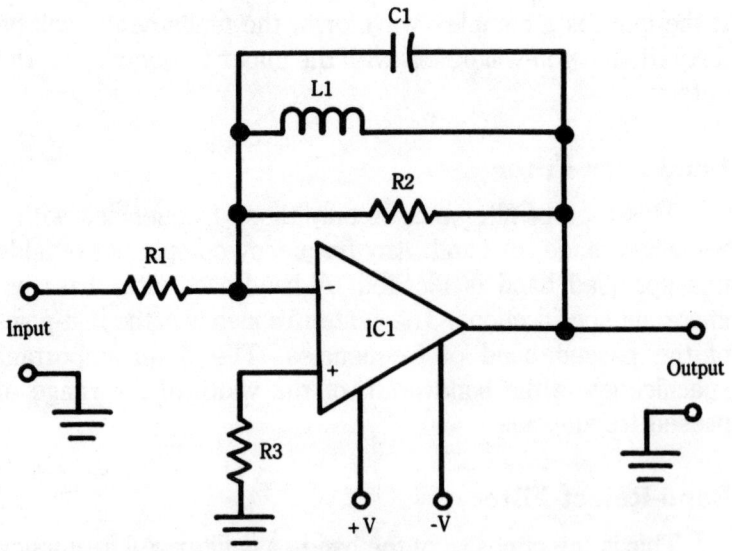

Fig. 7-7. This is a practical active band-pass filter. Project 40

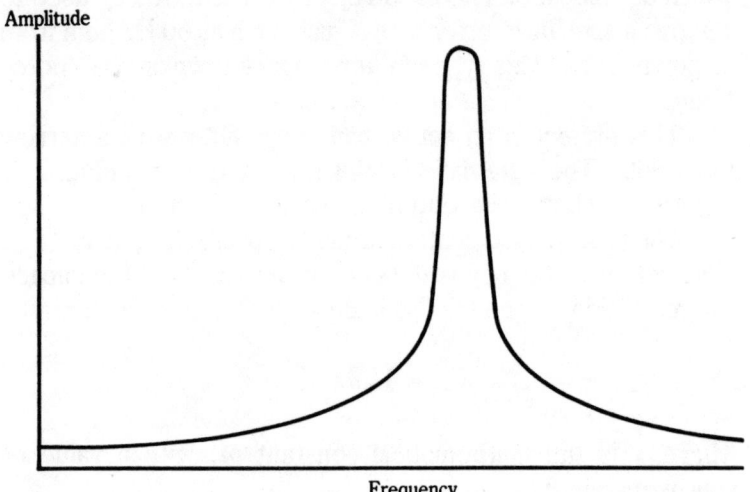

Fig. 7-8. This is the output graph for the circuit of Fig. 7-7.

Op amp filter circuits are often designed for unity gain at the center frequency (F_c). For unity gain, all the resistors in this circuit should have equal values.

To design for a specific desired center frequency, choose

a likely value for the coil (L1) and rearrange the frequency equation to solve for the necessary value of C1.

As an example, let's say we need the center frequency to be around 1500 Hz (1.5 kHz). If we use a 2.5 mH (0.0025 henry) coil for L1, the value of C1 should be:

$$\begin{aligned} C1 &= 1/(4\pi^2 \, F^2 \, L) \\ &= 1/(4 \times 3.14^2 \times 1500^2 \times 0.0025) \\ &= 1/(4 \times 9.8596 \times 2250000 \times 0.0025) \\ &= 1/221841 \\ &= 0.0000045 \text{ farad} \\ &= 4.5 \, \mu F \end{aligned}$$

Unless your application is very critical, a 5 μF capacitor should be close enough. Remember that capacitor tolerances tend to be rather wide.

A complete parts list for this project is given in Table 7-2. If you are going to be using this filter for serious audio applications, you should consider using a low noise op amp instead of the 741 specified in the parts list.

Table 7-2. Parts List for the Narrow Band-pass Filter Project of Fig. 7-7.

Schematic Label	Part
IC1	op amp (741, or similar)
L1	2.5 mH coil*
C1	5 μF capacitor (non-polarized)*
R1–R3	10K, ¼-watt resistor

*frequency determining component

PROJECT NO. 41: WIDE-BAND FILTER

Figure 7-9 shows a two stage active band-pass filter with a fairly wide pass-band. Multiple stage filters have steeper cut-off slopes than single stage filters. The steepness of the cut-off at either end of the pass-band is the chief advantage of this circuit.

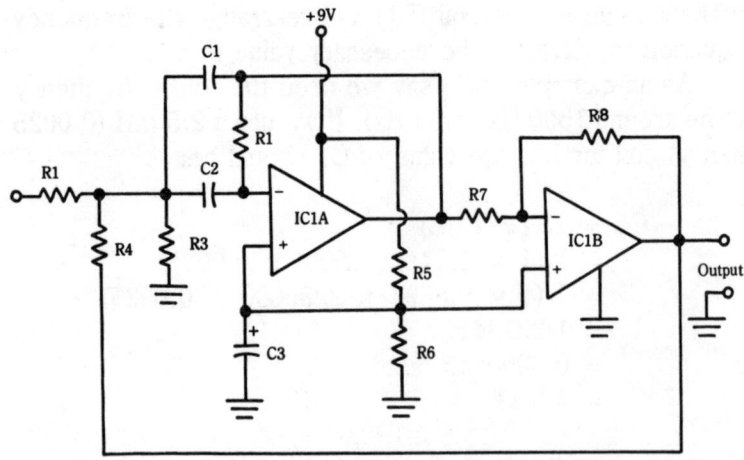

Fig. 7-9. This is a wideband band-pass filter circuit. Project 41

Both op amps in this circuit are part of a single IC. The 4136 is used. This chip contains four independent op amps in a single housing, so you could build two of these circuits around a single IC.

Using the component values listed in Table 7-3 will give a center frequency of about 1000 Hz for this filter circuit.

**Table 7-3. Parts List for the
Two-Stage Band-pass Filter Project of Fig. 7-9.**

Schematic Label	Part
IC1	4136 quad op amp (two sections used)
C1, C2	0.01 µF capacitor
C3	10 µF, 15V electrolytic capacitor
R1, R2	390K, ¼-watt resistor
R3	620 ohm, ¼-watt resistor
R4	620K, ¼-watt resistor
R5, R6	100K, ¼-watt resistor
R7	39K, ¼-watt resistor
R8	120K, ¼-watt resistor

PROJECT NO. 42: NOTCH FILTER

A notch filter, or band-reject filter is used to remove a specific range of interfering frequencies from a signal. A passive band-reject filter circuit is shown in Fig. 7-10. This is called a twin

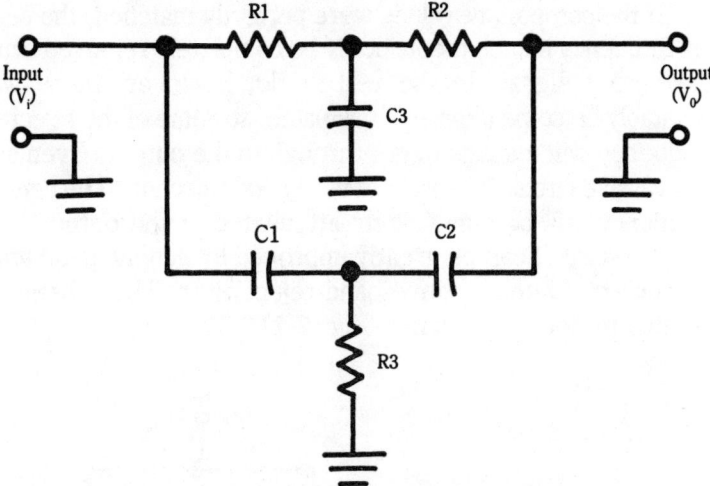

Fig. 7-10. This is a twin-T passive band-reject filter network.

T filter, because the components are arranged in two "T" patterns. One "T" is made up of R1, R2, and C3. The other "T" is made up of C1, C2, and R3.

Solving for the component values is not particularly difficult, because certain simple rules of proportion must be maintained for the circuit to work. The rules are:

$$R2 = R1$$
$$R3 = R1/2$$
$$C1 = C2$$
$$C3 = 2 \times C1$$

The center frequency, the mid-point of the rejected range, is sometimes called the null frequency. Whatever you want to call it, it can be determined with the following formula:

$$Fc = 1/(2\pi\, R1 C1)$$

Π, of course, is *pi*, the mathematical constant with a value of approximately 3.14. The formula can be rewritten as:

$$F_o = 1/(6.28 R1 C1)$$

If the component values were perfectly matched, the center frequency should theoretically be completely removed from the output signal. In the real world, however, there will inevitably be some degree of mismatch, so some of the rejected frequency will manage to get through to the output. Even so, this passive circuit is capable of very good rejection. The center frequency will be significantly attenuated at the output.

The circuit can be greatly improved by adding an op amp to convert it into an active band-reject filter. The schematic for this project is shown in Fig. 7-11.

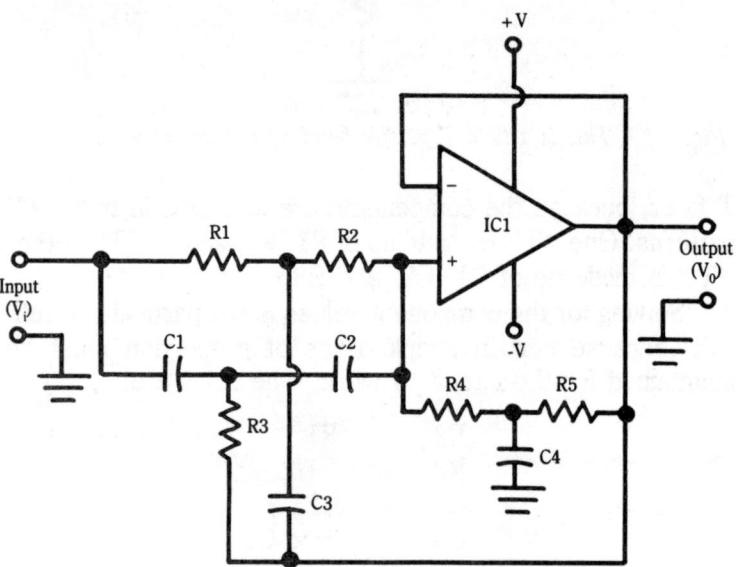

Fig. 7-11. This is an active "notch" filter circuit. Project 42

Resistors R1, R2, and R3, and capacitors C1, C2, C3, make up a passive twin T band-reject filter, just like the one discussed above. The center frequency equation is exactly the same.

A 741 op amp can be used for non-critical applications. For serious audio applications you might want to substitute a high grade low noise op amp IC. Remember to check the pin numbering.

Besides the op amp itself, three new components have been added to the circuit. They are R4, R5, and C4. Their values depend on the other component values in the circuit, and the desired Q (reciprocal of the bandwidth), according to the following equations:

$$R4 = R5 = 2QR1$$
$$C4 = C1/Q$$

A typical parts list for this project is in Table 7-4. With the component values listed here, the center frequency will be about 120 Hz, with a Q of 6. If you want a notch filter at a different center frequency, you can use the equations given above. To find the Q you can use this formula:

$$Q = F_c/BW$$

where F_c is the center frequency, and BW is the desired bandwidth. When I designed this circuit, I decided to use a narrow bandwidth of just 20 Hz, so the Q worked out to:

$$Q = 120/20 = 6$$

Table 7-4. Parts List for the Notch Filter Project of Fig. 7-11.

Schematic Label	Part
IC1	op amp (741, or similar)
R1, R2	12K, ¼-watt resistor
R3	6.2K, ¼-watt resistor
R4, R5	150K, ¼-watt resistor
C1, C2	0.1 µF capacitor
C3	0.2 µF capacitor
C4	0.015 µF capacitor

These components are for a 20 Hz notch, centered around 120 Hz, and a Q of 6. For different specifications, see text.

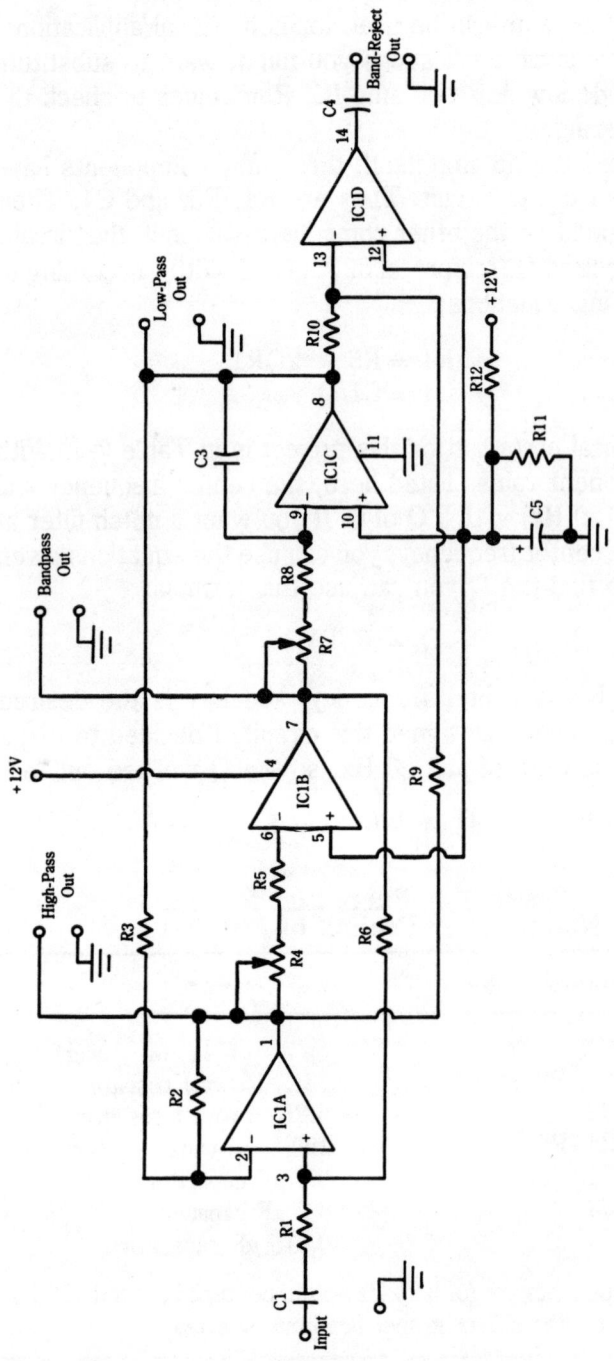

Fig. 7-12. A state variable filter has multiple outputs with different frequency responses. Project 43

PROJECT NO. 43: STATE VARIABLE FILTER

A state variable filter is sort of a "super filter." Most filters have a single output, however, a state variable filter has multiple outputs. Each of the outputs corresponds to one of the four basic filter types:

- low-pass
- high-pass
- band-pass
- band-reject

The circuit for a state variable filter project is shown in Fig. 7-12. The circuit is designed around the four stages of a single 4136 quad op amp IC.

Potentiometers R4 and R7 are used to set the cut-off frequencies of the low-pass and high-pass sections. Using the component values listed in Table 7-5, the cut-off frequency can be adjusted from about 300 Hz up to nearly 3000 Hz. For the best results, the potentiometers should have a reverse log type taper. If you cannot find potentiometers of this rather specialized type, you can use any potentiometer of the proper value, but the circuit will be harder to tune to a specific cut-off frequency.

Table 7-5. Parts List for the State Filter Project of Fig. 7-12.

Schematic Label	Part
IC1	4136 quad op amp
R1–R3, R6	100K, ¼-watt resistor
R9–R12	100K, ¼-watt resistor
R4, R7	500K potentiometer
R5, R8	39K, ¼-watt resistor
C1	0.01 μF capacitor
C2, C3	0.001 μF capacitor
C4	0.1 μF capacitor
C5	10 μF, 25V electrolytic capacitor

8

Miscellaneous Projects

This final chapter features a variety of interesting projects that don't quite fit into any of the categories of the earlier chapters.

PROJECT NO. 42: LED FLASHER

In terms of parts count, this is the simplest project in this book. In fact, it is probably one of the simplest circuits you'll ever come across. This LED flasher project is made up of just three components. In addition to the IC, which qualifies the project for inclusion here, only an LED and a capacitor are used.

The IC in this project is the LM3909, a low frequency oscillator. The LM3909 is specifically designed for LED flasher applications. Its function is simply to turn an LED on and off at a rate determined by the external capacitor. The pinout diagram for this simple eight pin IC is shown in Fig. 8-1.

The LM3909 has flexible power requirements. The supply voltage can be anything from 1.15 volts up to 6.0 volts. The low current drain of this circuit, typically under 0.5 mA (0.0005 amp), makes it very suitable for battery powered operation. The project can be powered from a single 1.5 volt cell. You can use any size battery, of course, but a small AA, or AAA cell is probably the most logical choice.

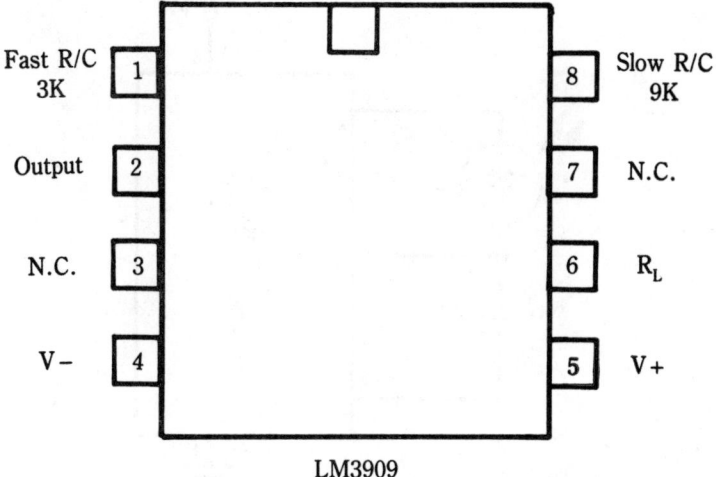

Fig. 8-1. The LM3909 is designed for LED flasher applications.

The circuit for this simple project is shown in Fig. 8-2. The LM3909 puts out pulses of about 2 volts to the LED, even if the supply voltage is only 1.5 volts.

The oscillator circuit in the LM3909 is designed to be self-starting. No external trigger signal is required.

The parts list, as short as it is, is given in Table 8-1. Be sure to experiment with other values of capacitor C1 for other flash rates.

PROJECT NO. 43: DUAL LED FLASHER

I'm not quite sure why it is, but LED flasher projects are great favorites among hobbyists (including myself). These circuits don't really do much in practical terms, but they're fascinating to watch. If you insist on finding a practical application, LED flashers make great eye catching displays.

When you get bored with watching a single LED blink on and off, you might want to move on to a two LED system, as in this project. The schematic is shown in Fig. 8-3.

The circuit is built around three stages of a quad Nor gate. Actually, all three gates are wired as inverters, so you could use three stages of a quad NAND gate, or half of a hex inverter IC.

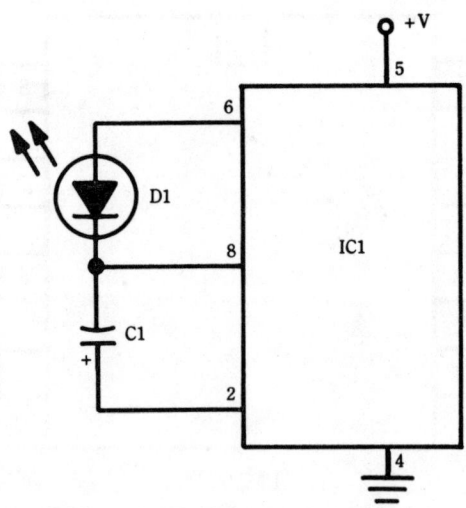

Fig. 8-2. This is a very simple LED flasher circuit. Project 44

Table 8-1. Parts List for the LED Flasher Project of Fig. 8-2.

Schematic Label	Part
IC1	LM3909 LED flasher IC
D1	LED
C1	1 μF electrolytic capacitor—experiment with other values

The first two stages are connected to form a simple oscillator. The frequency is determined by R2 and C1. The output of this oscillator drives LED1 directly, causing it to flash on and off.

The output of the oscillator is also fed into the third gate which inverts the signal—changing High signals to Low and Low signals to High. LED2 is driven by the output of this inverter. The signal fed to LED2 will always be at the opposite state of the signal fed to LED1. When LED1 goes on, LED2 goes off, and vice versa.

Experiment with different values for the frequency determining components, R2 and C1. If you set up too high a frequency, both LEDs will appear to be continuously lit.

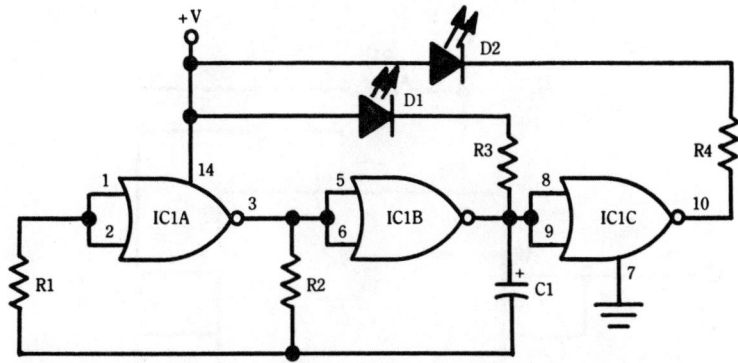

Fig. 8-3. This is a dual LED flasher circuit. Project 45

A suggested parts list for this project is given in Table 8-2.

PROJECT NO. 44: FREQUENCY DOUBLER

Sometimes it is necessary to change the frequency of an ac signal. It isn't much problem to divide the frequency down to a lower value, at least if you need to divide by a factor of two. Frequency division can be accomplished with one or more flip flops or bistable multivibrators. Each stage in series divides the signal by two.

But what if you need to increase the signal frequency? Figure 8-4 shows the circuit for a simple, inexpensive frequency doubler project. The parts list for this project, short as it is, is given in Table 8-3.

Table 8-2. Parts List for the Dual LED Flasher Project of Fig. 8-3.

Schematic Label	Part
IC1	CD4001 quad NOR gate
C1	10 µF, 35 volt electrolytic capacitor*
R1	1 Megohm, ¼-watt resistor
R2	180K, ¼-watt resistor*
R3, R4	330 ohm, ¼-watt resistor
D1, D2	LED

*frequency determining component

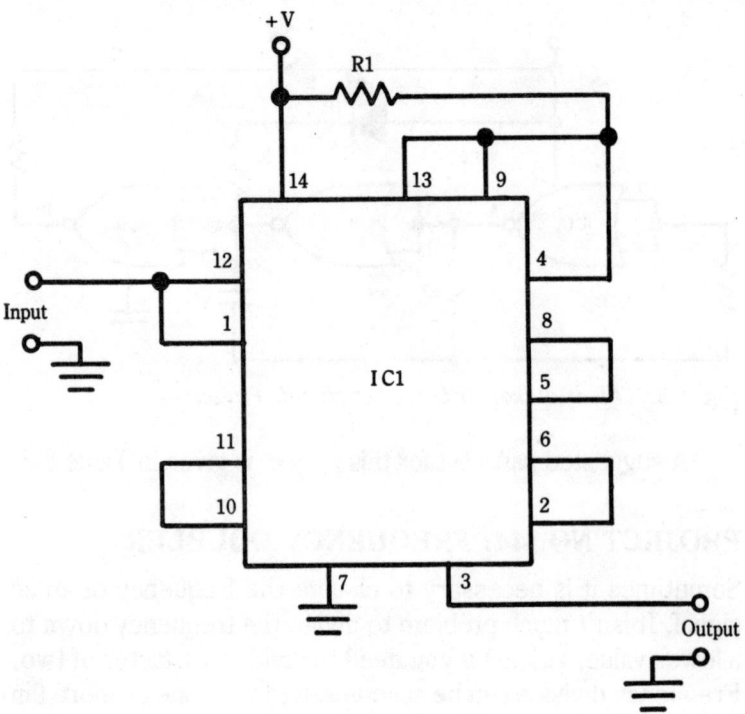

Fig. 8-4. The input frequency is doubled at the output of this circuit. Project 46

Table 8-3. Parts List for the Frequency Doubler of Fig. 8-4.

Schematic Label	Part
IC1	74C86 Exclusive OR gate
R1	1K, ¼-watt resistor

The output of this circuit will have exactly twice the frequency of the input signal. The output will always be a rectangular, square, or pulse wave, regardless of the input waveform.

Theoretically, any input waveform may be used with this project. But since this is a digital circuit, some analog waveforms might result in erratic operation. If you run into problems in your application, you might want to feed the input signal through a Schmitt trigger before feeding it into this circuit. This will only be necessary for some waveshapes, but it will never hurt.

The diagram is redrawn in Fig. 8-5 showing the logic gates individually.

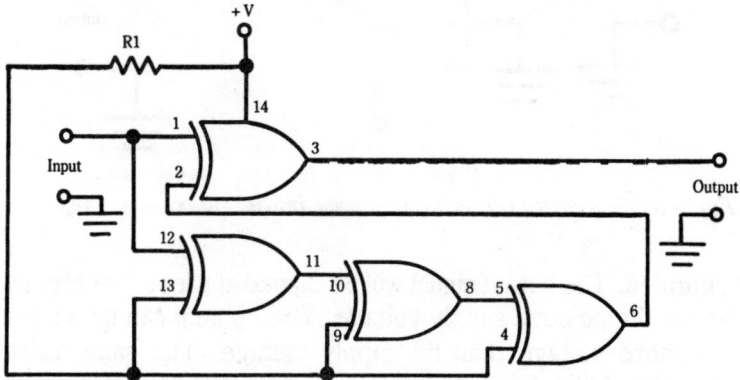

Fig. 8-5. The circuit of Fig. 8-4 is redrawn here to illustrate its logic functions.

PROJECT NO. 45: SINE-TO-SQUARE-WAVE CONVERTER

In some applications it is necessary to convert one waveform to another. The simple circuit shown in Fig. 8-6 is a sine-wave-to-square-wave converter. This circuit has just a single component—an op amp. No external passive components are needed.

The input signal does not have to be a sine wave. Almost any symmetrical ac signal at the input will produce a square wave at the output.

This converter circuit takes advantage of the extremely high (theoretically infinite) gain of the op amp. If the input signal is zero, the output signal will be zero, regardless of the gain, ignoring the effects of any internal offset voltages within the op amp, of course. But as soon as the input signal is even slightly higher than zero (positive), it will be amplified so much that the output of the op amp will almost instantly become

141

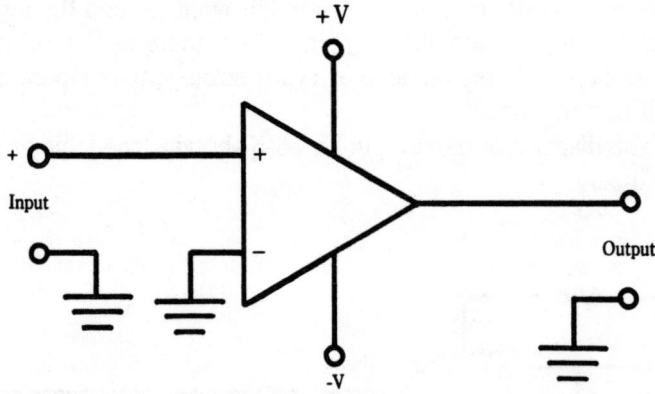

Fig. 8-6. This circuit converts a sine wave into a square wave. Project 47

saturated. The output signal will be clipped at a level just slightly below the positive supply voltage. The op amp can never put out more voltage than its supply voltage. The same thing happens if the input signal goes even slightly below zero (negative). In this case, the output is clipped at a level just above the negative supply voltage.

For any ac input signal, the output will switch back and forth between V+ and V− with very little transition time between the two extreme output states. If the input is a symmetrical waveform, like a sine wave, the output will be a neat square wave, as illustrated in Fig. 8-7. For non-symmetrical waveforms, the output will be a rectangular wave with a duty cycle other than 1:2, like one of the waveforms shown in Fig. 8-8.

This simple circuit can be used in electronic music synthesizers to create new sounds and tonal qualities. A sine wave, for example, has no harmonics, but a square wave or rectangular wave is very rich in harmonics. It could be interesting to run an instrument like an electric guitar through a circuit like this, especially if the squared off signal is mixed together with the original signal.

The squarer circuit can also be very useful for automated switching applications. A non-square waveform can be rather ambiguous and potentially confusing to control circuitry. The

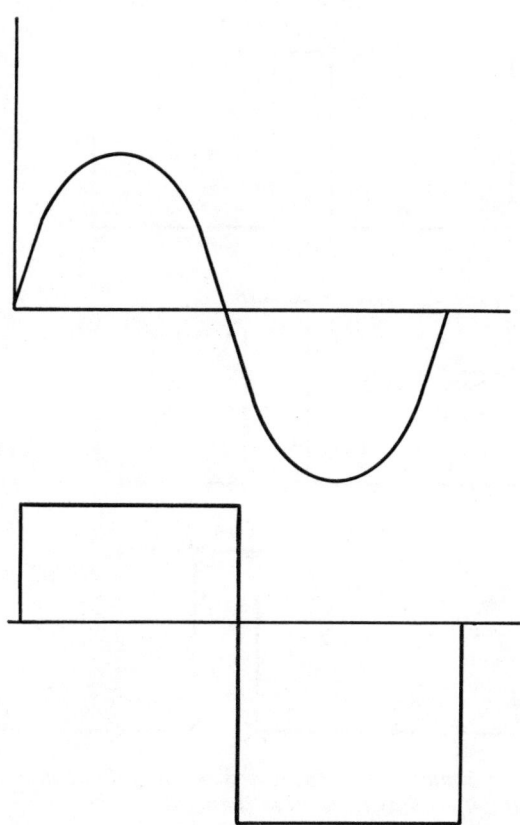

Fig. 8-7. If the input is a sine wave, the output will be a symmetrical square wave.

converter project will also allow you to interface an analog signal with digital circuitry. The supply voltages should be selected so that the two output states will be suitable for the logic family you are using in the digital portion of the circuit.

While not a true Schmitt trigger, this circuit can be used in many similar applications.

Since this project uses only a single component, there is no need for a parts list.

PROJECT NO. 46: AM TRANSMITTER

This is one of the more complex projects in this book, but I think you'll find it worth the extra effort and expense. It is

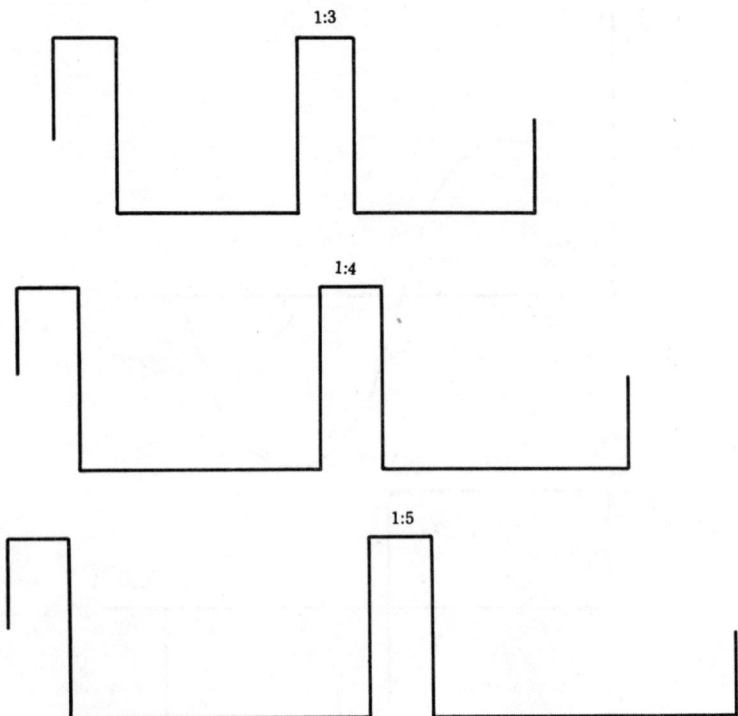

Fig. 8-8. Non-symmetrical input waveforms will result in a rectangular wave output with a duty cycle other than 1:2.

still a fairly easy project, and the total cost shouldn't add up to much more than $10, or perhaps even less.

With this project you can transmit your voice, or a control signal over any standard AM radio. The potential applications are almost endless.

Don't modify this circuit unless you know what you're doing. Increasing the output power could subject you to some heavy fines if you're caught by the FCC. It is legal to build a small unlicensed transmitter for the AM broadcast band (54 kHz to 160 kHz), but there are some very definite limitations. The transmitted signal may not cover more than a fifty foot range.

The schematic for this project is shown in Fig. 8-9, with the parts list appearing as Table 8-4. As you can see, this project requires a few more components than most of the other

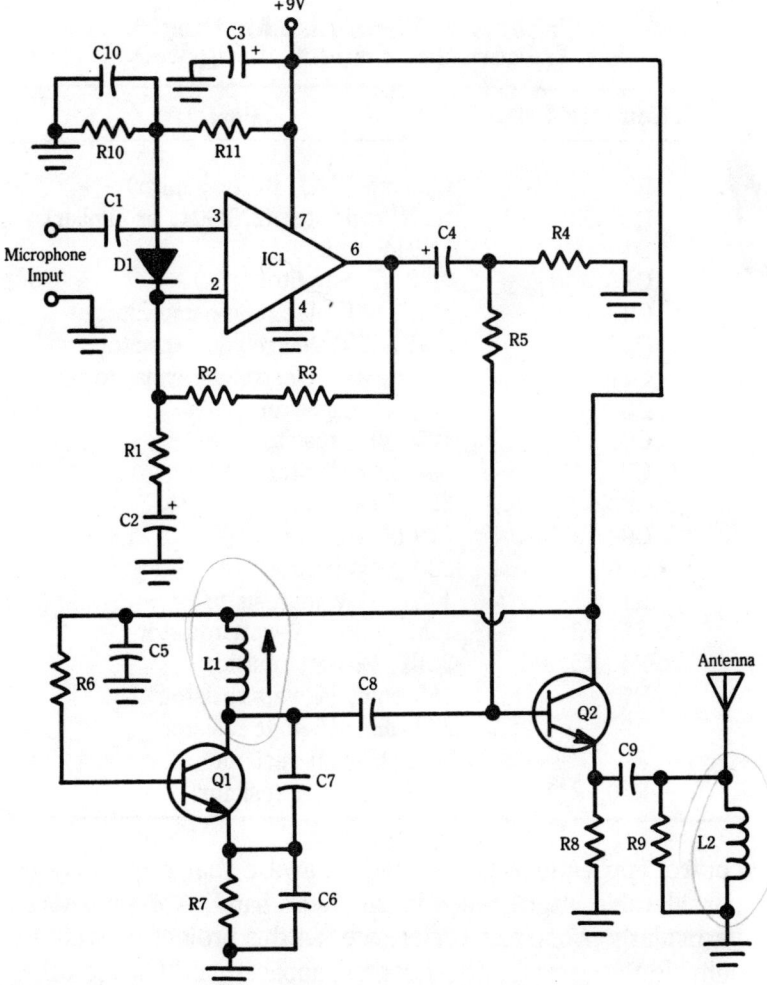

Fig. 8-9. This circuit allows you to transmit signals over almost any AM radio. Project 48

projects in this book, but the circuit is still centered around a single IC.

This circuit accepts an audio input signal. For voice use, you would use a microphone as the input device. Alternatively, you could use the transmitter to send wireless control signals. The control signal generator may be connected directly to the input of the transmitter. If you intend to use this project for

Table 8-4. Parts List for the AM Transmitter Project of Fig. 8-9.

Schematic Label	Part
IC1	Op amp (CA3140, or similar)
Q1, Q2	NPN transistor (2N3904, or similar)
D1	1N914 diode
C1	0.22 μF capacitor
C2	1 μF, 35V electrolytic capacitor
C3	5 μF, 35V electrolytic capacitor
C4	10 μF, 35V electrolytic capacitor
C5	0.05 μF capacitor
C6	1000 pF capacitor
C7	500 pF capacitor
C8	33 pF capacitor
C9	270 pF capacitor
C10	330 pF capacitor
R1	4.7K, ¼-watt resistor
R2, R3	1 Megohm, ¼-watt resistor
R4, R5, R6	2.2K, ¼-watt resistor
R7	560 ohm, ¼-watt resistor
R8	270 ohm, ¼-watt resistor
R9	33K, ¼-watt resistor
R10, R11	100K, ¼-watt resistor

control applications you should be aware that there may be considerable interference in this radio band. AM signals are particularly prone to interference, so this project may not be suitable for certain critical control applications. On the other hand, for many non-critical applications, this project is a fine, inexpensive method of wireless remote control.

PROJECT NO. 47:
BINARY-TO-HEXADECIMAL CONVERTER

Human beings generally count using a numbering system with a base of ten. There are ten distinct digits:

$$0 - 1 - 2 - 3 - 4 - 5 - 6 - 7 - 8 - 9$$

If a value larger than the largest available digit, 9 in this case, must be expressed, additional columns must be added. Each new column to the left increases its digit value by a factor equal to the base (ten) over its neighbor to the immediate right. For example:

$$253 = 2 \times (10 \times 10) + 5 \times 10 + 3 \times 1$$

We are so used to the base ten system, we don't even have to think about it. Other numbering systems are also possible. For example, digital circuitry is designed to use the binary numbering system. That is, there are just two digits:

$$0-1$$

The binary system is not used because circuit designers wanted to make things more difficult. A two digit system is the easiest to represent electronically without any ambiguity. No signal represents a "0." A present signal indicates a "1."

In the binary system, a value greater than one requires more than one column. A comparison of the binary and digital numbering systems is given in Table 8-5.

Table 8-5. Binary Equivalents of Decimal Numbers.

Decimal	Binary	Decimal	Binary
0	0000	9	1001
1	0001	10	1010
2	0010	11	1011
3	0011	12	1100
4	0100	13	1101
5	0101	14	1110
6	0110	15	1111
7	0111	16	10000
8	1000		

Notice that in the binary system, it is customary to add leading zeros.

As you can see, the binary system can quickly become very unwieldy for human beings to deal with. For example, the binary equivalent of decimal 1297 is 010100010001. It would be very, very easy to make a mistake copying a number like that. Unfortunately, it is not easy to convert back and forth between the binary and decimal numbering systems. To make life a little easier for people working with digital circuits, intermediate numbering systems are used, specifically the octal and the hexadecimal numbering systems. The octal numbering system has a base of eight. Each octal digit corresponds to three binary digits. The hexadecimal system uses sixteen digits. Letters from A to F are used to represent digit values greater than nine. Each hexadecimal digit represents four binary digits. A comparison of the decimal, binary, and hexadecimal numbering systems is given in Table 8-6.

Table 8-6. Binary and Hexidecimal Equivalents of Decimal Numbers.

Decimal	Binary	Hexadecimal
0	0000 0000	00
1	0000 0001	01
2	0000 0010	02
3	0000 0011	03
4	0000 0100	04
5	0000 0101	05
6	0000 0110	06
7	0000 0111	07
8	0000 1000	08
9	0000 1001	09
10	0000 1010	0A
11	0000 1011	0B
12	0000 1100	0C
13	0000 1101	0D
14	0000 1110	0E
15	0000 1111	0F
16	0001 0000	10
17	0001 0001	11
18	0001 0010	12

19	0001 0011	13
20	0001 0100	14
21	0001 0101	15
22	0001 0110	16
23	0001 0111	17
24	0001 1000	18
25	0001 1001	19
26	0001 1010	1A
27	0001 1011	1B
28	0001 1100	1C
29	0001 1101	1D
30	0001 1110	1E
31	0001 1111	1F
32	0010 0000	20
33	0010 0001	21
34	0010 0010	22
35	0010 0011	23

Returning to our earlier example, the large binary number can be broken up into four bit "bytes" (a binary digit is called a "bit"), and converted directly into hexadecimal:

```
0101    0001    0001
  5       1       1
```

or 511, in hexadecimal. That is a lot easier to remember and copy.

Now, finally we get to our project, a binary to hexadecimal converter. This circuit, illustrated in Fig. 8-10, accepts a four bit binary input and activates one of sixteen outputs.

No parts list is required for this project, since it consists of a single component—the IC. In this project we are using a 74C154 1-of-16 demultiplexer IC.

This project can be used to digitally control almost anything. Up to 16 external switching devices may be independently controlled by feeding the appropriate binary value into the inputs.

PROJECT NO. 48: D/A CONVERTER

It is often necessary, or at least desirable, to use a digital signal to operate some analog circuit. Some means of translating the digital numerical values into continuously varying analog

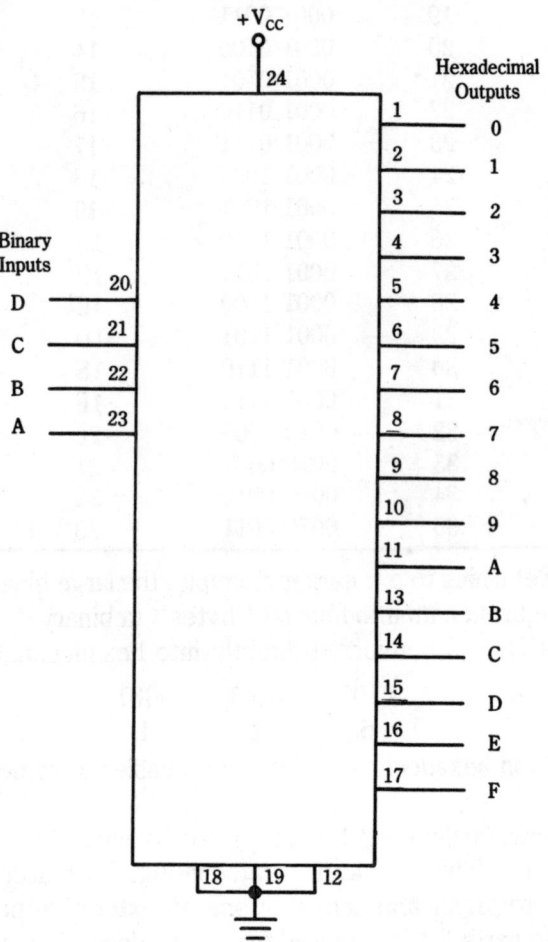

Fig. 8-10. This circuit converts binary values into their hexadecimal equivalents. Project 49

voltages or currents is needed. A circuit that performs such translation tasks is called a digital-to-analog converter, or D/A converter for short.

Op amps are frequently used in D/A conversion applications. A simple op amp D/A converter is shown in Fig. 8-11. This circuit is really nothing more than a summing amplifier. The input resistors are scaled to weight the inputs. The 2 (2) column should have twice the weighting of the 2 (1) column. The 2 (4) column should have twice the weighting

150

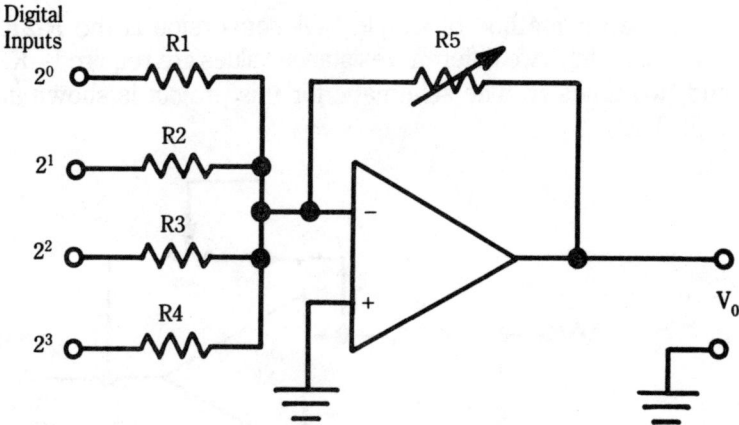

Fig. 8-11. This is a simple D/A converter circuit.

of the 2 (2) column, and four times the weighting as the 2 (1) column, and so forth. This can be extended as far as needed to allow one input resistor for each binary digit, or bit. Digital information is fed into the circuit in parallel fashion. Binary numbers with more than 8 digits (bits) would be awkward to handle for this simple circuit.

High quality precision resistors with very low tolerances should be used in this application. Standard 5 percent resistors probably won't be close enough for any but the least critical applications.

The problem with this circuit is that each successive resistor must have half the value of its predecessor, and twice the value of the next input resistor. This results in an unwieldy range of resistance values if more than four or five bits are to be used. In addition, standard resistor values are not based on multiples of two. The repeated halving, or doubling, results in some oddball values. For example, a four bit version of this circuit might use the following resistance values:

$$
\begin{aligned}
R1 &= 100K \\
R2 &= 50K \\
R3 &= 25K \\
R4 &= 12.5K
\end{aligned}
$$

A better method of simple D/A conversion is the R-2R approach. Only two different resistance values are required—R, and two times R. The schematic for this project is shown in Fig. 8-12.

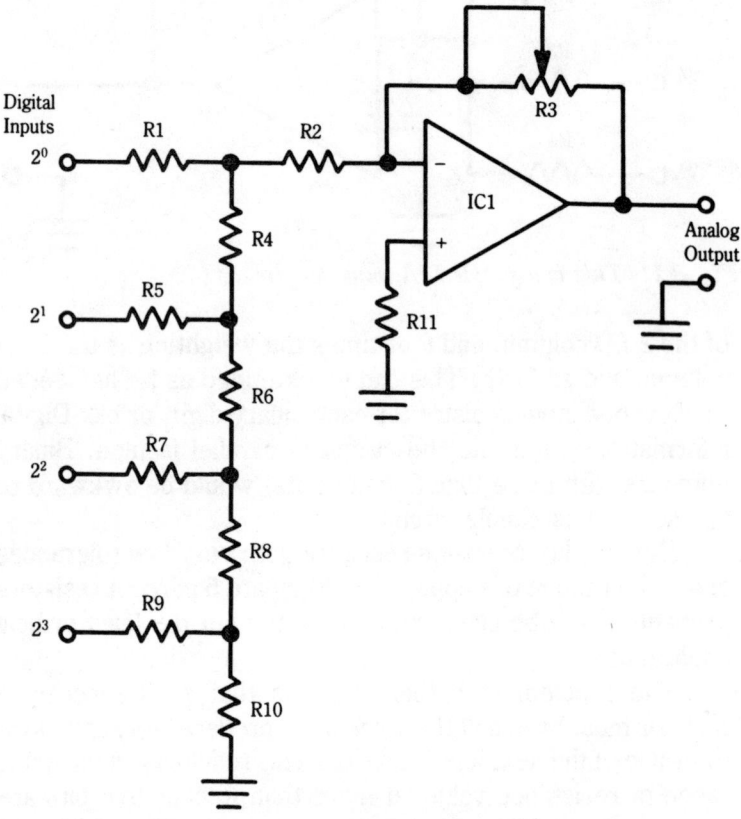

Fig. 8-12. A better D/A converter can be built using the R-2R network. Bonus Project.

For this circuit, the following relationships must be maintained between the resistance values:

$$R4 = R6 = R8 = R11 = R$$
$$R1 = R2 = R5 = R7 = R9 = R10 = 2R$$

Feedback resistor R3 sets the overall circuit gain. Except for R3, all the resistors in this circuit have one of the two basic values—R, or 2R. Ten K is a good choice for R, making 2R equal to 20K. The easiest way to find the 20K resistors is to make them from two 10K resistors in series. If you do not use high precision low tolerance resistors, you should match the resistances with an ohmmeter before using them in this circuit.

While only four bit inputs are shown here, the R-2R ladder can be extended almost infinitely to accommodate as many input bits as your application requires. Larger binary numbers with more bits offer finer resolution:

Range	No. of Bits	Maximum Count (decimal)
4	0000 - 1111	15
8	00000000 - 11111111	255
16	0000 0000 0000 0000 - 1111 1111 1111 1111	65,535

The system cannot really be expanded infinitely. Eventually the resistances presented to the Least Significant Bits (LSBs) will become too large for a reasonable amount of signal to get through.

The voltage available at each bit input is equal. If the logic value of that bit is 0, the input is effectively at ground potential. For a logic 1, a constant positive voltage level is applied to the bit input. The most significant bit (MSB) passes through the smallest number of resistors; therefore, it presents the strongest voltage at the op amp's input. The LSB voltage has to pass through the maximum number of resistors, so it will experience the maximum voltage drop, and present the weakest signal at the op amp's input. Each logic 1 bit presents a proportionately weighted voltage to the op amp. Any bits that are at logic 0 are ignored. The output will be an analog voltage which is directly proportional to the digital values being presented at the inputs to the converter.

A typical parts list for this project appears in Table 8-7.

PROJECT NO. 49: FREQUENCY HALVER
Sometimes we need to lower the frequency of a signal. The simplest type of division is by two. The output frequency of this circuit is exactly one half of the input frequency.

Table 8-7. Parts List for the D/A Converter Project of Fig. 8-12.

Schematic Label	Part
IC1	op amp (741, or similar)
R1, R2, R5 R7, R9, R10	20K, ¼-watt resistor (two 10K resistors in series)
R4, R6, R8, R11	10K, ¼-watt resistor
R3	25K potentiometer (gain control—may be replaced with a fixed resistor)

The circuit for this project is shown in Fig. 8-13. Since no external components beyond the IC are used, there is no need for a parts list.

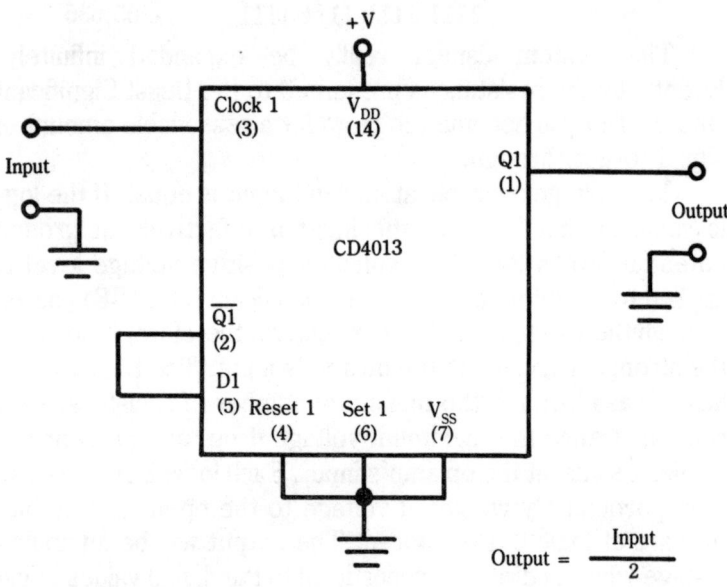

Fig. 8-13. A D-type flip flop can be used to divide an input signal by two. Bonus Project

The IC used here is the CD4013 dual D-type flip flop. The pinout diagram for this device is shown in Fig. 8-14. Notice that we are only using half of this chip. Two divide-by-two circuits can be hooked up with a single IC.

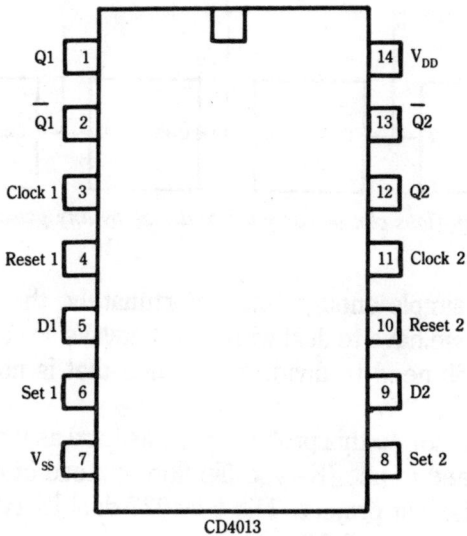

Fig. 8-14. The CD4013 is a dual D-type flip-flop IC.

This circuit is certainly simple enough. It is simply a toggled flip flop. Each time the flip flop is triggered by an input pulse, its output reverses state. Since the state must be changed twice for a complete cycle, it follows that there must be two complete input cycles for every complete output cycle. There will be twice as many input pulses as there are output pulses. In other words, the output frequency is one half the input frequency.

In musical applications, dividing the frequency by two drops the pitch one octave.

BONUS PROJECT: DIVIDE-BY-THREE CIRCUIT

In project No. 49 we divided a signal frequency by two. But what if our application requires us to divide the frequency by some other value?

If the desired division value is a power of two (4, 8, 16, etc.), there is no problem. We can simply cascade divide-by-two counters, as illustrated in Fig. 8-15. The first divider divides by two. The second counter divides the first one's output by two. This is the same as dividing by 2×2, or four. This can be repeated by as many stages as you need.

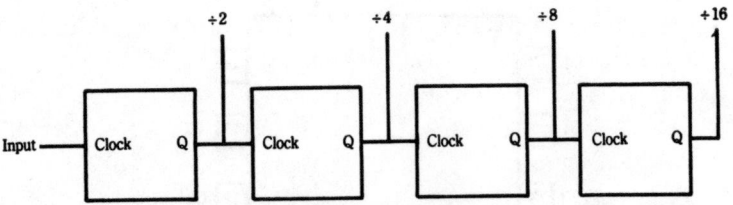

Fig. 8-15. Flip flops can be cascaded to divide by any power of two.

This is simple enough, but unfortunately, the real world applications we have to deal with aren't always so cooperative. Often we will need to divide by a value that is not a power of two.

The solution to this problem isn't as hard as it may seem. First, we need to use JK-type flip flop, instead of the D-type flip flop of the last project. The CD4027 dual JK-type flip flop IC is shown in Fig. 8-16.

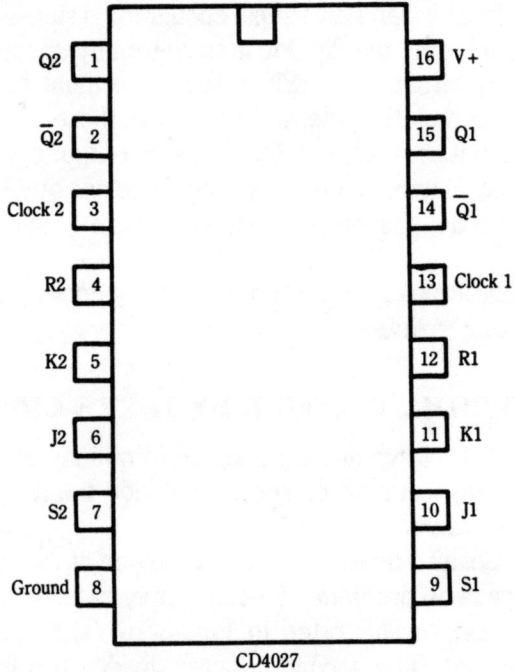

Fig. 8-16. The CD4027 is a dual JK-type flip flop IC.

Cascade enough JK-type flip flop stages to divide by the next higher power of two. For example, to divide by three we need to jump up to the next higher power of two, which is four. This requires two stages. Then we simply force a reset after the desired count (three, in this case). A divide by three circuit is shown in Fig. 8-17. To give you a better idea of how this project works, the circuit is redrawn in functional form in Fig. 8-18.

Once again, since this project uses a single IC (CD4027) and no external components, a parts list is unnecessary.

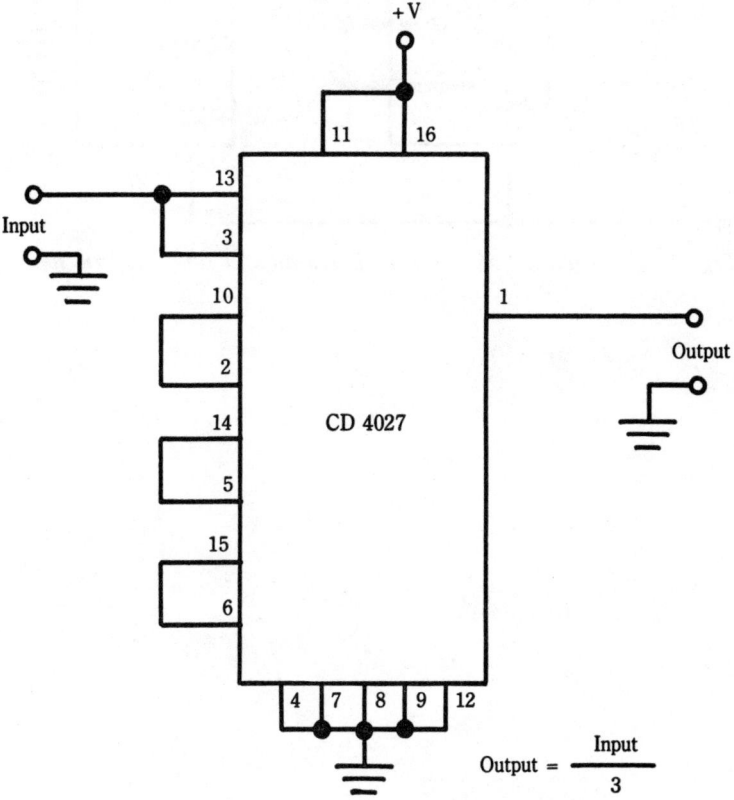

Fig. 8-17. This is a divide by three circuit. Bonus Project

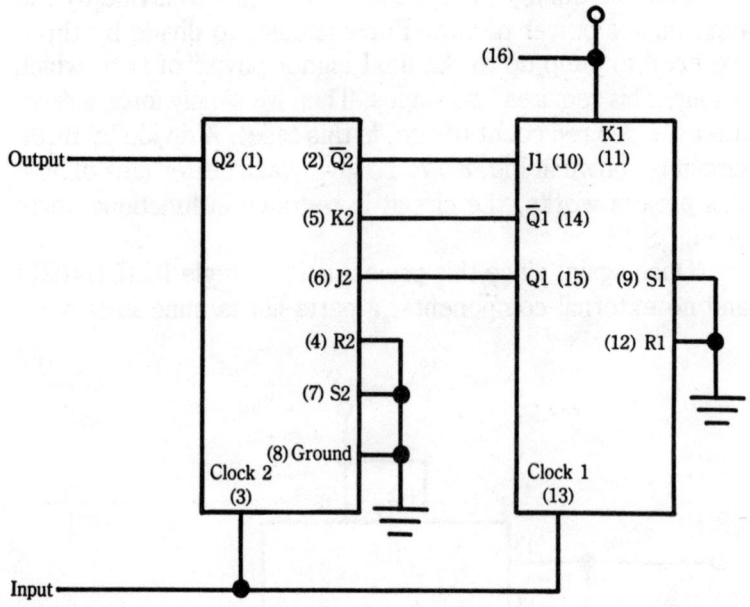

Fig. 8-18. The circuit of Fig. 8-17 is redrawn in function form here.

Index

A
active band pass filter, 125, 127
active band reject filter, 127
active high pass filter, 126
active low pass filter, 125
alarms, 112
   burglar, 112
   flooding, 115
   lights-off, 118
   lights-on, 117
AM transmitter, 143
amplifiers, 48-67
   audio, 49
   audio mixer, 55
   ceramic phono, 54
   difference, 59
   digital linear, 57
   logarithmic, 62
   signal splitter, 55
audible continuity tester, 96
audio amplifier, 49
audio function generator, 89
audio mixer, 55
automated guest greeter, 25

B
band pass filter, 122
   active, 125, 127
band reject filter, 122
   active, 127
bargraph, 98
binary to hexadecimal converter, 146
bounce, correcting switch, 31
breadboarding, 11-13, 14
burglar alarm, 112

C
capacitance, 11
capacitors
   parallel configurations for, 11
   ratings and codes for, 11
   series configurations for, 11
   substitution values for, 6, 10
ceramic phono amplifier, 54
circuit modules, 12
circuits,
   divide by three, 155
   high-frequency, 15

indicator, 112
integrated, 14-16
switching, 17-47
codes, for capacitors, 11
comparator, voltage, 100
components
grab bag, 3
heat-sensitive, 14
locating and purchasing, 1-5
mail-order, 2
mounting of, 14
organization and storage of, 4
orientation of, 15
recycled, 3
substitutions for, 5-11
construction techniques, 13-14
continuity tester, audible, 96
converter
binary to hexadecimal, 146
D/A, 149
sine to square wave, 141

D

D/A converter, 149
debouncer, switch, 31
detector
light range, 118
voltage range, 103
difference amplifier, 59
digital linear amplifier, 57
digital phase-shift oscillator, 88
digital sine wave generator, 82
divide by three circuit, 155
doubler, frequency, 139
dual LED flasher, 137

E

effective resistance, 8-10
errors, 12

F

555 triangular wave generator, 76
feedback, 127
filters, 122-135
active band pass, 125-127
active band reject, 127
active high pass, 126
active low pass, 125
band pass, 122
band reject, 122
high pass, 122
integrator, 123
low pass, 122
notch, 131
state variable, 135
wide-band, 129
flasher
dual LED, 137
LED, 136
flooding alarm, 115
frequency doubler, 139
frequency halver, 153
function generator, audio, 89
fuses, substitution values for, 6

G

gain
band-reject filter, 127
positive vs. negative, 48
generator,
audio function, 89
digital sine wave, 82
op amp square wave, 68
programmable square wave, 73
sound-pocket, 91
tone-burst, 93
555 triangular wave, 76
grab bag components, 3
guest greeter, automated, 25

H

halver, frequency, 153
heatsinking, 14, 15
high-frequency circuits, 15
high pass filter, 122
   active, 126

I

indicator circuits, 112
integrated circuits
   heat-sensitivity of, 14
   orientation of, 15
   pin #1 marking on, 16
   soldering of, 15
integrator, 123

L

LED flasher, 136
   dual, 137
light range detector, 118
light-activated relay, 19
lights-off alarm, 118
lights-on alarm, 117
linear amplifier, digital, 57
logarithmic amplifier, 62
logic probe, 107
long-duration timer, 41
low pass filter, 122
   active, 125

M

mail-order components, 2
measuring devices, 96
miscellaneous projects, 136
mixer, audio, 55
mounting components, 14

N

negative gain, 48
notch filter, 131

O

op amp sine wave oscillator, 76
op amp square wave generator, 68
orientation of components, 15
oscillator, 12, 68
   digital phase-shift, 88
   op amp sine-wave, 76

P

pc boards, 14
phase-shift oscillator, digital, 88
phono amplifier, ceramic, 54
positive gain, 48
power supplies, 12
precautions, 14-16
programmable square wave generator, 73
programmable timer, 35

R

range detector
   light, 118
   voltage, 103
ratings,
   for capacitors, 11
   for resistor tolerance, 7, 8
recycling components, 3
relay
   light-activated, 19
   telephone-activated, 24
   voice-activated, 21
   vox, 21
resistance, effective, 8, 9, 10
resistors
   parallel configurations for, 8, 9, 10
   series configurations for, 8
   substitution values for, 5, 6
   tolerance ratings for, 7, 8

S

safety, 14-16
Schmitt trigger, 28
semiconductors, substitution values for, 6
signal generators, 68
  555 triangular wave generator, 76
  audio function generator, 89
  digital phase shift oscillator, 88
  digital sine wave generator, 82
  op am sine wave oscillator, 76
  op amp square-wave generator, 68
  programmable square-wave generator, 73
  sound-pocket generator, 91
  tone-burst generator, 93
signal splitter, 55
sine to square wave converter, 141
sine wave generator, digital, 82
sine wave oscillator, op amp, 76
sockets, 14
  solderless, 11
soldering, 14, 15
solderless sockets, 11
sound-pocket generator, 91
splitter, signal, 55
square wave generator
  op amp, 68
  programmable, 73
state variable filter, 135
substituting components, 5-11
switch
  touch, 17
  touch, timed, 46
switch debouncer, 31
switching circuits, 17-47
  automated guest greeter, 25
  light-activated relay, 19
  long-duration timer, 41
  programmable timer, 35
  Schmitt trigger, 28
  switch debouncer, 31
  telephone-activated relay, 24
  timed touch switch, 46
  touch switch, 17
  twenty-four hour timer, 38
  vox relay, 21

T

telephone-activated relay, 24
test equipment, 96
timed touch switch, 46
timer
  long-duration, 41
  programmable, 35
  twenty-four hour, 38
tolerance ratings, resistors, 7, 8
tone-burst generator, 93
touch switch, 17
  timed, 46
transmitter, AM, 143
triangular wave generator, 76
twenty-four hour timer, 38

U

universal pc boards, 14

V

values,
  substituting capacitor, 6, 10
  substituting resistor, 5, 6
voice-activated relay, 21
voltage comparator, 100
voltage range detector, 103
vox relay, 21

W

wide-band filter, 129